D1520746

Springer Series in
OPTICAL SCIENCES 112

founded by H.K.V. Lotsch

Springer Series in
OPTICAL SCIENCES

The Springer Series in Optical Sciences, under the leadership of Editor-in-Chief *William T. Rhodes*, Georgia Institute of Technology, USA, provides an expanding selection of research monographs in all major areas of optics: lasers and quantum optics, ultrafast phenomena, optical spectroscopy techniques, optoelectronics, quantum information, information optics, applied laser technology, industrial applications, and other topics of contemporary interest.
With this broad coverage of topics, the series is of use to all research scientists and engineers who need up-to-date reference books.

The editors encourage prospective authors to correspond with them in advance of submitting a manuscript. Submission of manuscripts should be made to the Editor-in-Chief or one of the Editors. See also http://www.springer.de/phys/books/optical_science/

A. Goetzberger V.U. Hoffmann

Photovoltaic Solar Energy Generation

With 138 Figures

Professor Dr. Adolf Goetzberger
Dipl.-Wirt Volker U. Hoffmann
Fraunhofer ISE, Heidenhofstr. 2, 79110 Freiburg
E-mail: goetzb@ise.fhg.de

ISSN 0342-4111

ISBN 3-540-23676-7 Springer Berlin Heidelberg New York

Library of Congress Control Number: 2004116389

Springer is a part of Springer Science+Business Media.

springeronline.com

Printed in Germany

Typesetting and prodcution: PTP-Berlin, Protago-TEX-Production GmbH, Berlin
Cover concept by eStudio Calamar Steinen using a background picture from The Optics Project. Courtesy of John T. Foley, Professor, Department of Physics and Astronomy, Mississippi State University, USA.
Cover production: *design & production* GmbH, Heidelberg

Printed on acid-free paper SPIN: 10899342 57/3141/YU 5 4 3 2 1 0

Preface

The intention of this book is to provide an impression of all aspects of photovoltaics (PV). It is not just about physics and technology or systems, but it looks beyond that at the entire environment in which PV is embedded. The first chapter is intended as an introduction to the subject. It can also be considered an executive summary. Chapters 2–4 describe very briefly the basic physics and technology of the solar cell. The silicon cell is the vehicle for this description because it is the best understood solar cell and also has the greatest practical importance. A reader who is not interested in the physical details of the solar cell can skip Chap. 2 and still understand the rest of the book. In general, it was the intention of the authors to keep the book at a level that does not require too much previous knowledge of photovoltaics. Chapter 5 is devoted to other materials and new concepts presently under development or consideration. It intends to provide an impression of the many possibilities that exist for the conversion of solar radiation into electricity by solid state devices. These new concepts will keep researchers occupied for decades to come. Chapter 6 gives an introduction to cell and module technology and also informs the reader about the environmental compatibility and recycling of modules.

The following chapters are devoted to practical applications. Chapters 7 and 8 introduce systems technology for different applications. The environmental impact of PV systems and their reliability is the subject of Chap. 9. It is pointed out that PV systems, in particular, modules belong to the most durable industrial products today. Systems efficiency is explained in Chap. 10. In particular, performance ratio, which permits comparison of systems independent of location, is introduced. In Chap. 11, we turn to questions of market and costs. Although PV is the most expensive renewable energy source today, it has a large cost reduction potential. Future cost can be predicted by referring to the learning curve that links cost to cumulated production. In order to realize this potential, PV needs long-term reliable public support. This support can occur in many different ways, as is shown in Chap. 11. The experience gained so far indicates that feed-in tariffs are the best market support mechanism.

Chapter 12 contains a vision of the future of PV. Decentralized systems in buildings, etc. have the best short- and medium-term prospects, but large-scale PV power plants are also a possibility for the more distant future. On the other hand, PV is not the only renewable energy source, and in the future other such sources will be competing for markets. These other sources and how they compare to PV are discussed in Chap. 13. In Chap. 14, finally, frequently encountered arguments against PV are answered by referring to information provided in previous chapters of this book. In this manner, the summary and conclusion are combined in a somewhat unconventional way.

Freiburg, January 2005 *Adolf Goetzberger, Volker Hoffmann*

Contents

1 What Is Photovoltaics?

1.1 What Is Photovoltaics?

Photovoltaics (abbreviated PV) is the most direct way to convert solar radiation into electricity and is based on the photovoltaic effect, which was first observed by Henri Becquerel [1] in 1839. It is quite generally defined as the emergence of an electric voltage between two electrodes attached to a solid or liquid system upon shining light onto this system. Practically all photovoltaic devices incorporate a pn junction in a semiconductor across which the photovoltage is developed (see Chap. 2). These devices are also known as solar cells. Light absorption occurs in a semiconductor material. The semiconductor material has to be able to absorb a large part of the solar spectrum. Dependent on the absorption properties of the material, the light is absorbed in a region more or less close to the surface. When light quanta are absorbed, electron hole pairs are generated, and if their recombination is prevented they can reach the junction where they are separated by an electric field. Even for a weakly absorbing semiconductor like silicon, most carriers are generated near the surface. This leads to the typical solar cell structure of Fig. 1.1.

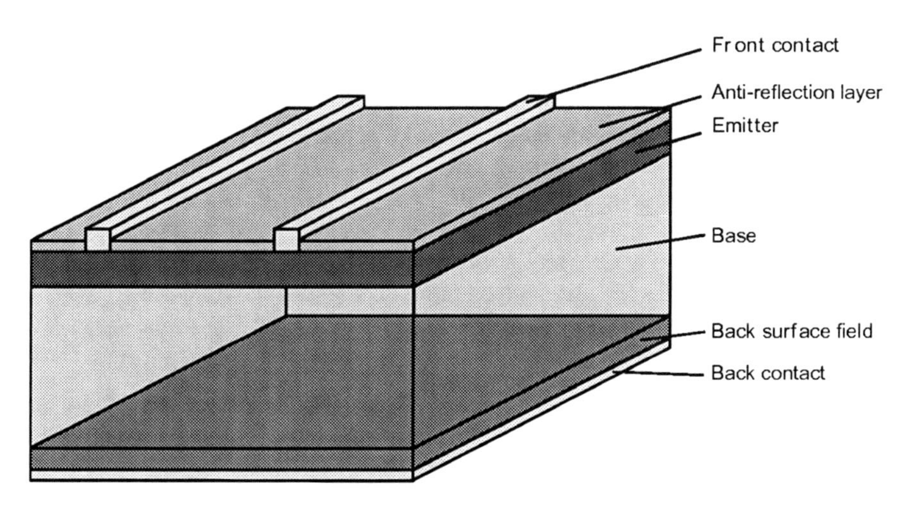

Fig. 1.1. Typical solar cell structure

The pn junction that separates the emitter and base layer is very close to the surface in order to have a high collection probability for the photogenerated charge carriers. The thin emitter layer above the junction has a relatively high resistance which requires a well designed contact grid, also shown in the figure.

For practical use solar cells are packaged into modules containing either a number of crystalline Si cells connected in series or a layer of thin-film material which is also internally series connected. The module serves two purposes: It protects the solar cells from the ambient and it delivers a higher voltage than a single cell, which develops only a voltage of less than 1 Volt. The conversion efficiencies of today's production cells are in the range of 13 to 16%, but module efficiencies are somewhat lower. The best laboratory efficiency of crystalline silicon achieved so far is 24.7%, which approaches the theoretical limit of this type of solar cell.

As we shall see in Chap. 5, pn junctions and semiconductors are not the only way to achieve photovoltaic conversion. The future may hold many new materials and concepts.

1.2 Short History of Photovoltaics

1.2.1 Technology

The photovoltaic effect remained a laboratory curiosity from 1839 until 1959, when the first silicon solar cell was developed at Bell Laboratories in 1954 by Chapin et al. [2]. It already had an efficiency of 6%, which was rapidly increased to 10%. The main application for many years was in space vehicle power supplies.

Terrestrial application of photovoltaics (PV) developed very slowly. Nevertheless, PV fascinated not only the researchers, but also the general public. Its strong points are:

- direct conversion of solar radiation into electricity,
- no mechanical moving parts, no noise,
- no high temperatures,
- no pollution,
- PV modules have a very long lifetime,
- the energy source, the sun, is free, ubiquitous, and inexhaustible,
- PV is a very flexible energy source, its power ranging from microwatts to megawatts.

Solar cell technology benefited greatly from the high standard of silicon technology developed originally for transistors and later for integrated circuits This applied also to the quality and availability of single crystal silicon of high perfection. In the first years, only Czochralski (Cz) grown single crystals were used for solar cells. (For a description of the Czochralski technique, see

Sect. 3.2.2). This material still plays an important role. As the cost of silicon is a significant proportion of the cost of a solar cell, great efforts have been made to reduce these costs. One technology, which dates back to the 1970s, is block casting [3] which avoids the costly pulling process. Silicon is melted and poured into a square SiO/SiN-coated graphite crucible. Controlled cooling produces a polycrystalline silicon block with a large crystal grain structure (see Sect. 3.2.2).

From solid state physics we know that silicon is not the ideal material for photovoltaic conversion. It is a material with relatively low absorption of solar radiation, and, therefore, a thick layer of silicon is required for efficient absorption. Theoretically, this can be explained by the semiconductor band structure of silicon in which the valence band maximum is offset from the conduction band minimum (see Fig. 2.5). Since the basic process of light absorption is excitation of an electron from the valence to the conduction band, light absorption is impeded because it requires a change of momentum. The search for a more suitable material started almost with the beginning of solar cell technology. This search concentrated on the thin-film materials. They are characterized by a direct band structure, which gives them very strong light absorption.

Today, the goal is still elusive, although promising materials and technologies are beginning to emerge. The first material to appear was amorphous Silicon (a-Si). It is remarkable that even the second contender in this field is based on the element silicon, this time in its amorphous form. Amorphous silicon has properties fundamentally different from crystalline silicon. However, it took quite some time before the basic properties of the material were understood. The high expectancy in this material was curbed by the relatively low efficiency obtained so far and by the initial light-induced degradation for this kind of solar cell (so-called Staebler–Wronski effect) [4]. Today, a-Si has its fixed place in consumer applications, mainly for indoor use. After understanding and partly solving the problems of light-induced degradation, amorphous silicon begins to enter the power market. Stabilized cell efficiencies reach 13%. Module efficiencies are in the 6–8% range. The visual appearance of thin-film modules makes them attractive for facade applications.

Beyond amorphous silicon there are many other potential solar cell materials fulfilling the requirement of high light absorption and are therefore suitable for thin-film solar cells. They belong to the class of compound semiconductors like GaAs or InP, which are III–V compounds according to their position in the periodic table. Other important groups are II–VI and I–III–VI_2 compounds, which, just like the elemental semiconductors, have four bonds per atom. It is clear that an almost infinite number of compounds could be considered. From the mostly empirical search only very few promising materials have resulted. Foremost are Copper Indium Diselenide (CIS) and Cadmium Telluride (CdTe). Already by the early 1960s cadmium sulfide/copper sulfide solar cells were under development [9]. Problems with low efficiency and insufficient stability prevented further penetration of this material.

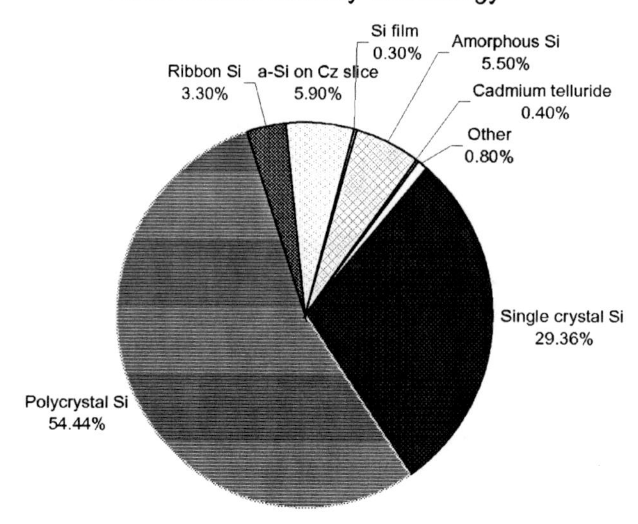

Fig. 1.2. Market shares of different technologies for 2002

The new technology is based on the ternary compound semiconductors $CuInSe_2$, $CuGaSe_2$, $CuInS_2$ and their multinary alloy $Cu(In,Ga)(S,Se)_2$ (in the following text: CIGS). The first results of single crystal work on $CuInSe_2$ (CIS) were extremely promising, but the complexity of the material looked complicated as a thin-film technology. Pioneering work, however, showed immediate success. It became evident that CIS process technology is very flexible with respect to process conditions. In later developments, the addition of Ga and S helped to increase the efficiency. The best laboratory efficiency has recently reached a remarkable 18.9%. CIS/CIGS modules are now available on the market in small quantities.

Thin-film solar cells based on CdTe have a very long tradition and are also just at the onset of commercial production. After a long and varied development phase, they arrived at cell efficiencies of 16% and large-area module efficiencies of over 10%.

In spite of the complicated manufacture and the high cost, crystalline silicon still dominates the market today and probably will continue to do so in the immediate future. This is mostly due to the fact that there is an abundant supply of silicon as raw material, high efficiencies are feasible, the ecological impact is low, and silicon in its crystalline form has practically no degradation. The market shares of different technologies in 2002 are shown in Fig. 1.2.

The various forms of crystalline silicon have together a share of 93%. Single crystal and cast poly material had about equal share for a long time. Recently, cast material has surpassed single crystals. Newer types of crystalline silicon like Ribbon and Si film are not yet very important. A newcomer is a-Si on crystalline silicon (see Chap. 5). Of the true thin-film materials thath are summarized as "others" amorphous silicon is dominant. As mentioned before, its market is mainly in consumer products. These market shares are rather stable and change only in an evolutionary manner. The dominance of the element silicon in its crystalline and amorphous forms is an overwhelming 99%. Of all the other materials only CdTe has a market share of only 0.4%.

1.2.2 Applications

The need to provide power for space vehicles provided an excellent point of entry for solar cells starting in the late 1950s. Solar cells can work reliably and without maintenance for long periods. This provided an opportunity for further development. Efficiency was increased and resistance against radiation was studied and improved. At the same time, PV energy supply systems for those very demanding conditions were developed.

In 1958, the first 108 solar cells for the supply of the Vanguard satellite were put into orbit. They performed even better than predicted and powered the satellite much longer than expected. The demand for solar cells climbed rapidly in the following years, leading to a small industrial production. The consequence was not only an improvement of the electrical parameters of the cells, but also a drop in prices. This, in turn, lead to a modest use of solar cells in terrestrial applications, but space remained the main market for more than a decade.

The breakthrough for terrestrial photovoltaics can be traced directly to the oilshock of 1973/74. Experts in all industrialized nations started to look for alternatives to the scarce and expensive mineral oil. They discovered photovoltaics and recognized a possible candidate for a future nonfossil energy supply. Newly emerging development institutions in the United States, Europe, and Asia occupied themselves not only with the development of cells, but also with systems and system components. The problems to be solved were formidable: The cost of PV energy had to be reduced by a factor of 1,000. This referred not only to the cells, but also to the entire system.

Since then, the price for grid-connected systems has been reduced by a factor 100. How was this accomplished? Obviously, an energy source as expensive as PV has no chance in an open market. Governments in some European countries, the U.S., and Japan initiated large support programs, because they were convinced of the great potential of photovoltaics. The success of cost reduction resulted from an interaction of several more or less coordinated initiatives: Development of better solar cells and systems, demonstration programs for testing and optimization of systems, and, finally, market support programs for grid-connected generators.

As a result, production expanded with remarkable growth rates between 20 and 40% per year with corresponding cost reductions. The most important demonstration and market support programs are:

- The German 1000 Roof Measurement and analysis program [5]. (For more information, see Chap. 11) Some results of the German 1,000 Roofs Program are published in [6].
- The 100,000 Roof Program in Germany.
- The 1 Million Roof Program in the U.S. (This program also comprises thermal systems.)
- The Italian Roof top Program [7].
- Smaller programs were introduced in Austria [8] and Switzerland [9].

Probably the most important tool for market development are the feed-in laws in several European countries and Japan. These laws provide a more or less adequate compensation for PV energy fed into the grid. For more information, see Chap. 11. In parallel, a fully economic market developed for consumer products and power supplies for remote installations.

Today, we have generators ranging from several milliwatts in consumer products to grid-connected systems in the kilowatt range up to central power plants of several megawatts.

1.3 Relevance of PV, Now and in the Future

Although the theoretical potential of PV worldwide is very high, it is difficult to quote a single figure for this potential. Of the total solar radiation reaching the earth's surface each year only a minute part (about 0.003%) is equivalent to global electricity demand today. The potential of PV is part of the potential of all kinds of utilization of solar radiation energy. In this respect, there is no realistic limitation for this potential. Compared to wind energy, which is another and presently more economical renewable electricity source, PV has the advantage that it is not limited to certain geographic locations. PV even today is in use practically everywhere. On the other hand, the amount of radiation depends on geography and climate, particularly on latitude. There is a difference of about a factor of 2.5 in radiative energy between the most arid desert regions and Central Europe. A serious problem in most locations is the intermittent nature of solar energy.

Examples of the daily and seasonal fluctuations of radiation are shown in Figs. 1.3 and 1.4. Figure 1.3 demonstrates the case of Freiburg in southern Germany, where we see a large but strongly fluctuating solar input in summer and very low availability in winter. Figure 1.4 in contrast gives an example of a very sunny desert climate, Khartoum, Sudan. Solar input is much more uniform on a daily and yearly scale. In the first case, seasonal storage is required for an all-solar system, while in Khartoum daily storage is sufficient.

Even in Central Europe, which is not blessed with an abundance of sunshine, a part of electricity demand (more than 50%) could be supplied by solar electricity in theory. In reality, many obstacles will have to be overcome before even a small percentage will be reached. In more northern (or southern) countries a big problem is the seasonal mismatch of supply and demand.

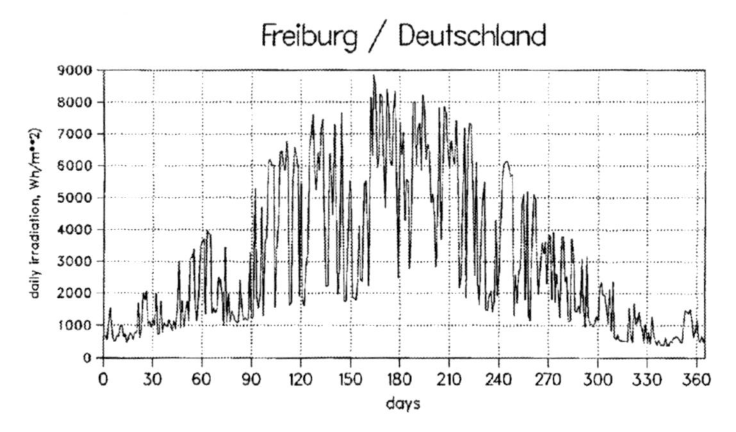

Fig. 1.3. Daily global radiation for one year in Freiburg, southern Germany

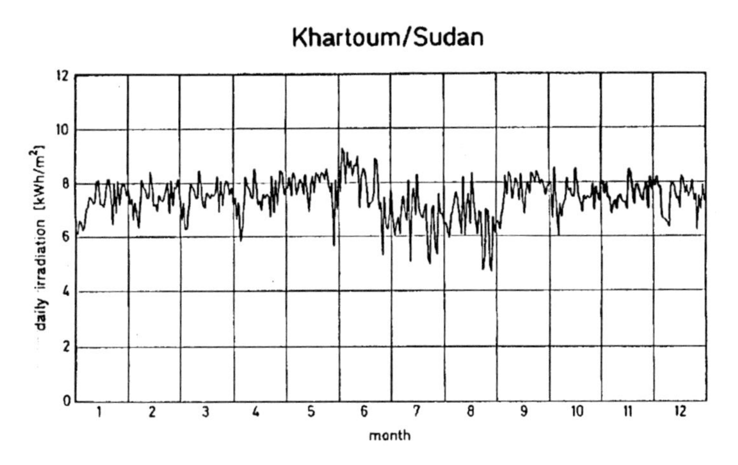

Fig. 1.4. Daily global radiation for one year in Khartoum, Sudan

Significant contributions can only be expected from grid-connected systems. In this case, the grid is used for storage. Until an economic way of seasonal energy storage becomes available, the practical limit may be about 10% of total generating capacity dependent on the elasticity of the grid. This is still a large amount of energy and very far from today's contribution. It should also be kept in mind that a combination of renewable energy sources with different stochastics like wind and PV provide a more even generating capacity.

The potential for Germany has been evaluated in several studies [10, 11]. In very approximate terms the result is that by using all suitable roof areas, about 20% of capacity could be reached. When comparing capacities, it should be realized that the continuous average power of a PV system is only about one tenth of peak power. Beyond roofs, other areas could also be used, like roads and rails, which could add the same amount. A still much bigger potential lies in unused agricultural areas. Further, it can be shown that by optimized mounting of PV generators the same land areas can be used simultaneously for PV and crop cultivation [12] (see also Sect. 9.4). Such high levels of PV generation, however, are not very likely in the foreseeable future because seasonal storage would be required.

In climatic zones with a higher and seasonally less variable solar radiation, very high contributions of PV-electricity are possible. It is obvious that the same solar cell if mounted in a desert area close to the equator would generate 2.0 to 2.5 times more electricity at correspondingly lower cost than in Europe. Arguments against this are the problem of intercontinental electricity transport and of security of supply. Nevertheless, it is conceivable that in a distant future, PV farms will be set up in desert areas and the energy will be transported to the consumers by long-distance grids or in the form of hydrogen.

PV today is economical only if it does not have to compete with grid electricity. Nevertheless, the technology is only at the beginning of its development and hopes are high for further large cost reductions. At present, however, it is not obvious that the cost of PV can reach present levels of the cost of base load electricity, but it can reach consumer retail prices. Besides development of technology, market expansion is a proven way of bringing down cost. In several countries that take their obligation to reduce greenhouse gases seriously, comprehensive support programs for distributed PV installations have been legislated. One example is the German renewable energy law, which stipulates that utilities have to pay for PV electricity fed into the grid about 0.5 €/kWh for twenty years. This reimbursement is reduced by 5% each year for new installations in order to stimulate cost reduction. The expected cost trends for the future are described in Chap. 12.

1.4 Markets, Economics

The main problem, as will be shown later, is the high cost of solar cells. Nevertheless, costs are dropping continuously and a remarkable market development has taken place. The photovoltaic world market in 2002 was more than 500 MW_p[1] per year, corresponding to a value of roughly US$ 1 billion. This is a remarkable market, but still far away from constituting a noticeable contribution to world energy consumption. Market growth in the last decade was between 15 and 25% per year and has risen recently to 30% and even 40%, as is shown in Fig. 1.5.

This market growth would be very satisfying if it could be maintained for ten to fifteen years. Cost of PV electricity would fall rapidly, as is outlined in later chapters. The main motivation for developing solar energy is the desire

[1] The power of cells and modules is measured in W_p (peak Watt), output power at a normalized radiation of about 1,000 W and a standardized spectral distribution and temperature (AM 1.5).

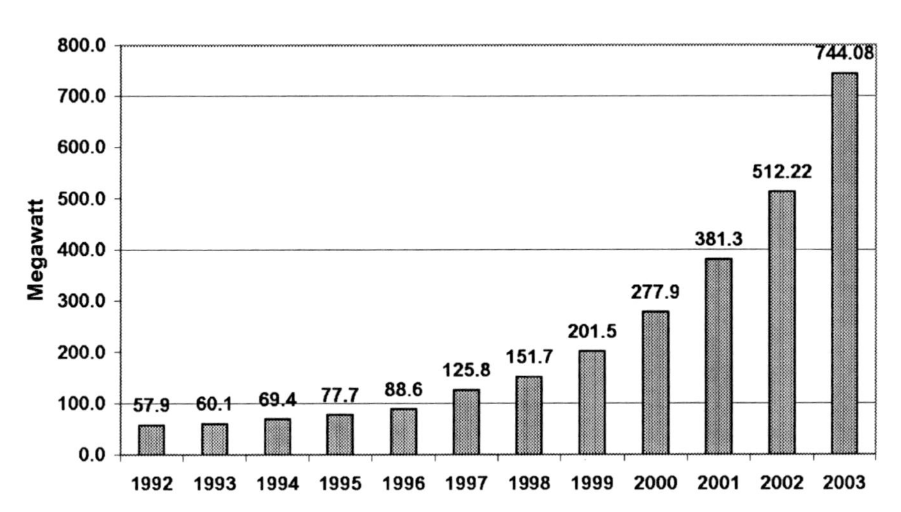

Fig. 1.5. Development of PV world markets in MW_{peak} (MW_{peak} is defined as power under full sun, approximately $1\,kW/m^2$)

to reduce dependence on depletable fossil fuels with their adverse effect on the environment.

There are two major market sectors, grid-connected and so-called stand-alone systems. The former delivers power directly to the grid. For this purpose the dc current from the solar modules is converted into ac by an inverter. The latter supplies power to decentralized systems and small-scale consumer products. A major market currently being developed is in solar home systems, supplying basic electricity demand of rural population in developing countries. The magnitude of this task can be appreciated if one is aware that about 2 billion persons are without access to electricity today. At present, both markets need subsidies, the grid-connected installations because PV is much more costly than grid electricity, and solar home systems because the potential users lack the investment capital. On the other hand, there is also a significant industrial stand-alone market that today is fully economical.

Because of its high potential, the market is hotly contested and new companies are entering constantly. It is significant that several large oil companies have now established firm footholds in photovoltaics. Indeed, a recent study of possible future energy scenarios up to the year 2060 published by the Shell company predicts a multigigawatt energy production by renewable energies, including photovoltaics. On the other hand, the strong competition leads to very low profit margins of most participants of this market.

Starting in 2000, the market showed an accelerated growth of more than 30%. There are good chances that this growth will continue for at least some years because some countries have adopted aggressive measures to stimulate the grid-connected market, as mentioned above. In order to meet the growing demand, many PV companies are in the process of setting up substantial new cell and module production capacities. The consequences this will have regarding the availability of semiconductor grade silicon will be discussed in Chap. 3.

2 Physics of Solar Cells

2.1 Basic Mechanisms of Energy Conversion

As we shall see in Chap. 5, different mechanisms and materials can be employed for the conversion of solar energy into electricity, but all practical devices, at least until today, are based on semiconductors.

Semiconductors are solids and, like metals, their electrical conductivity is based upon movable electrons. Ionic conductors are not considered here. The primary consideration here is the level of conductivity. Materials are known as

- conductors at a conductivity of $\sigma > 10^4\,(\Omega\text{cm})^{-1}$;
- semiconductors at a conductivity of $10^4 > \sigma > 10^{-8}\,(\Omega\text{cm})^{-1}$;
- non-conductors (insulators) at a conductivity of $\sigma < 10^{-8}\,(\Omega\text{cm})^{-1}$.

This simple categorization is, however, hardly an adequate criterion for a definition, and it is predominantly other characteristics, in particular the thermal behavior of conductivity, that form the basis for classification. This is where metals and semiconductors behave in an opposing manner. Whereas the conductivity of metals decreases with increasing temperature, in semiconductors it increases greatly. So what is a crystalline solid? At this point, we wish to differentiate between two separate categories. On the one hand, there are the so-called amorphous substances. In these, the structure of individual atoms and molecules displays almost no periodicity or regularity. Crystalline solids, on the other hand, are distinguished by a perfect (or near perfect) periodicity of atomic structure. These materials naturally make it much easier to understand the physical characteristics of solids. Therefore, the explanation of semiconductor characteristics and the physical principles of photovoltaics is normally based upon crystalline semiconductors, and in particular crystalline silicon.

In common with all elements of the fourth group of the periodic table, silicon has four valence electrons. These atoms are arranged in relation to each other such that each atom is an equal distance from four other atoms and that each electron forms a stable bond with two neighboring atoms. This type of lattice is known as the diamond lattice, because diamond – comprising of tetrahedral carbon – has this lattice structure. These bonds are extremely strong. This is demonstrated by other physical characteristics such as the hardness of these materials.

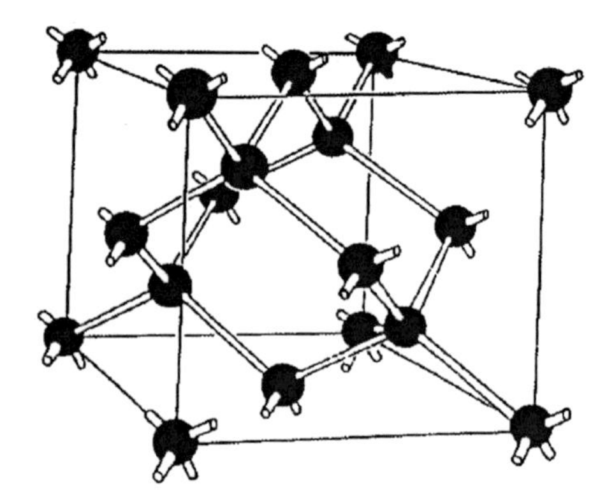

Fig. 2.1. The diamond lattice

Figure 2.1 shows the structure of a diamond lattice. We do not wish to go into further details about this structure at this point. Please refer to the specialist literature on solid state physics.

For electrical conductivity to occur in this type of crystal, some of these bonds must be broken. Clearly, this can only occur if energy is expended. At an absolute temperature of $T = 0\,\mathrm{K}$, no bonds are broken, i.e., no free electrons are present. At $T = 0\,\mathrm{K}$, the semiconductor is an insulator.

So what is the energy level structure in this type of crystal? We know that according to Bohr's theory of the atom, electrons in an isolated atom can only occupy well-defined energy levels. If we bring two or more atoms close together in an imaginary experiment, then an interactive effect will occur, splitting the energy levels of these bonded atoms. The number of frequencies increases with the number of atoms coupled. This leads to discrete energy bands with energy levels that can be occupied by electrons separated by gaps in which there can be no electrons. If we transfer this analogy to the interactive effect in a crystal lattice, then the splitting of energy levels can be represented as follows.

The vertical axis of Fig. 2.2 represents electron energy, and the horizontal axis represents the distance of atoms from one another. As the distance decreases, the energy levels of the atoms split up more and more. Similar to the mechanical analogy, the energy bands become increasingly broad. At a specified distance between atoms (d), it is clear that there is an energy gap between the two upper bands, the valence band and the conduction band, in which there can be no electrons. The energy gap is called the band gap, E_{g}, of a semiconductor. It further applies that when $T = 0\,\mathrm{K}$, because no bonds are broken, none of the energy levels in the outer conduction band will be occupied. In the valence band, however, all available energy levels are occupied. This means that no energy can be absorbed from an external electrical field, i.e., electric current cannot flow. The semiconductor is then

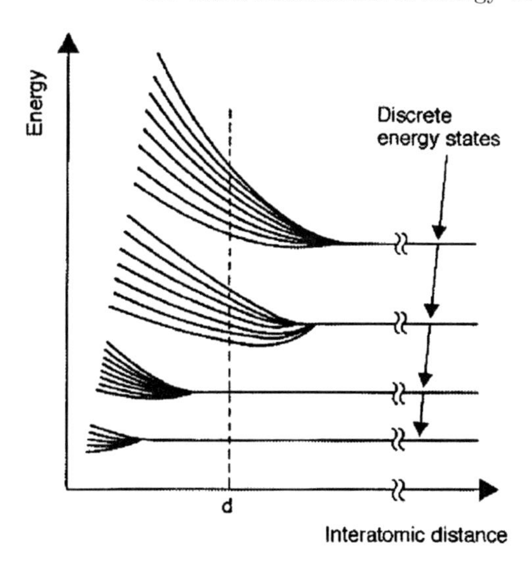

Fig. 2.2. Splitting of energy levels in a crystalline lattice

an insulator. Only at higher temperatures does it show conductivity, because then some electrons occupy the energy levels in the conduction band.

This band structure, with an energy gap between the two outer energy bands, also occurs in insulators. Semiconductors and insulators differ only in the size of the band gap. In a semiconductor, even at "normal temperatures" (e.g., at room temperature) some electrons can jump the band gap, thus giving rise to electrical conductivity. In insulators, the band distance is so large that at normal temperatures no electrons can jump the gap. Normal values for the energy of this band gap for semiconductors lie within the range of a few tenths of electron-volts to approximately 2 eV, whereas for insulators these energies are significantly higher.

Conduction can occur in a semiconductor in the following manner: In the broad temperature range of "normal" temperatures the conduction band is "almost empty" and the valence band "almost full" of electrons. "Almost empty" in the conduction band means that only a few electrons are in the permitted energy states. Although all these states lie near the edge of the band, there still are numerous unoccupied states close to the occupied levels, so these electrons are capable of reaching a higher level by an almost continuous process. This means that when connected to an electric field, energy can be continuously taken up. Therefore, electrons can move in the direction of an electric field. It is then possible to treat the conduction electrons as the electrons in metals are treated in classical physics. Owing to the level of dilution, they influence each other very little, but they are in a state of continuous close interaction with the lattice of the crystal. This interaction is highly complex and can only be considered statistically.

Analogous to this is the behavior in the "almost full" valence band. Some energy levels in this band are not occupied by electrons, and these energy levels also lie close to the edge of the valence band. As in the conduction band, these empty states are surrounded by numerous occupied states. This means that one such empty state can wander around within the valence band. This empty state is known as a hole or defect electron. It has proved sensible to treat this hole as an individual, i.e., as a charge carrier. It is evident that this charge carrier has a positive charge. The defect electron or hole, like the electron, is a second type of charge carrier. This is an extremely useful formalism for dealing with the phenomenon of conductivity in semiconductors.

The resistivity of pure silicon at $T = 300\,\mathrm{K}$ is extremely high, approximately 300,000 Ωcm. It also varies very significantly with temperature. There is, however, in addition to increasing the temperature, another highly effective method of altering the concentration of charge carriers in a semiconductor and thus its conductivity, namely by the purposeful introduction of certain impurity atoms into the crystal.

If we replace a silicon atom in the crystal structure with an element from the fifth group of the periodic table, for example (e.g., phosphorus), this atom "brings" five valence electrons with it. Only four of these electrons are required to bond to the crystal structure. It is therefore plausible that the fifth electron is relatively loosely bound and can, therefore, be "ionized" even at low temperatures. We call these elements from the fifth group of the periodic table "donors", as they can easily "donate" electrons. In addition to phosphorus, the elements arsenic and antimony are also used as donors in semiconductor technology. Because of the low activation energy of the extra electrons, they are ionized even at low temperature.

We can also use elements from the third group of the periodic table as dopants. The elements boron, aluminum, gallium, and indium are used in semiconductor technology. The missing bonding electron of a trivalent dopant atom leads to the creation of a hole and thus an increase in the positive conductivity of the semiconductor. This is therefore called a p-type conductor, and these types of dopant are known as acceptors. The holes are now dominant, i.e., the majority charge carriers, and the electrons are the minority charge carriers. The same regularity applies as in the case of doping with pentavalent atoms, i.e., even at low temperatures all holes are active. The energy levels of a number of chemical elements in the energy gap can be seen in Fig. 2.3. Donor levels are near the conduction band and acceptor levels are near the valence band. As can also be seen, some elements cause levels near the center of the gap. These levels are called recombination centers because they cause excess free electrons and holes to recombine across the gap. This recombination interferes with the operation of solar cells (and most other semiconductor devices) and should be avoided. Since even trace amounts of some impurities are very deleterious, semiconductors have to be extremely pure.

Next, we look at the generation of charge carriers by absorption of light in semiconductors. Unlike opaque metals, semiconductors display what is for

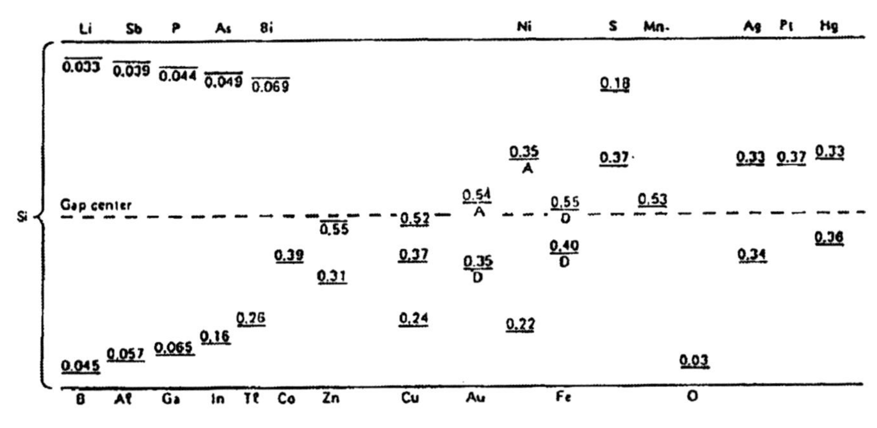

Fig. 2.3. Energy levels of chemical elements in the gap

them characteristic absorption behavior. The most important characteristic is the existence of the so-called absorption edge. For wavelengths λ, at which the photon energy ($E = \mathrm{h}c/\lambda$, where c is the speed of light in a vacuum and h is the Plank's constant) is greater than the energy of the forbidden band, light is, depending upon the thickness of the material, almost completely absorbed. In the case of long wavelength light, almost no absorption takes place due to its low energy. In this spectral region, the semiconductor is transparent. In the case of silicon, the band edge lies within infrared at $\lambda \sim 1.11\,\mu\mathrm{m}$. Therefore, silicon is excellently suited as a base material for infrared optics, but not as ideal for absorption of the solar spectrum, as we shall see later.

The intensity of the light entering the crystal is weakened during its passage through the crystal by absorption. The absorption rate is thus – as in many other cases of physical behavior – proportional to the intensity that is still present. This leads familiarly to an exponential reduction in intensity and can be described mathematically as follows:

$$F_x = F_{x,0} \exp\big(-\alpha_\lambda (x - x_0)\big)\,, \tag{2.1}$$

where F_x is the number of photons at point x; $F_{x,0}$ is the number of photons on the surface $x = 0$; and α_λ is the absorption coefficient.

The latter is itself dependent upon the wavelength, and determines the penetration depth of the light and therefore the thickness of crystal necessary to absorb most of the penetrating light. The absorption length x_L is also often introduced, corresponding to the value $x_\mathrm{L} = l/\alpha$. At this absorption length the intensity F_x is reduced to $(l/e)F_{x,0}$ (approximately 37%).

Absorption in semiconductors is a so-called basic lattice absorption, in which one electron is excited out of the valence band into the conduction band, leaving a hole in the valence band. Certain peculiarities of this process should be taken into account. A photon possesses a comparatively large amount of energy, but according to the De Broglie relationship

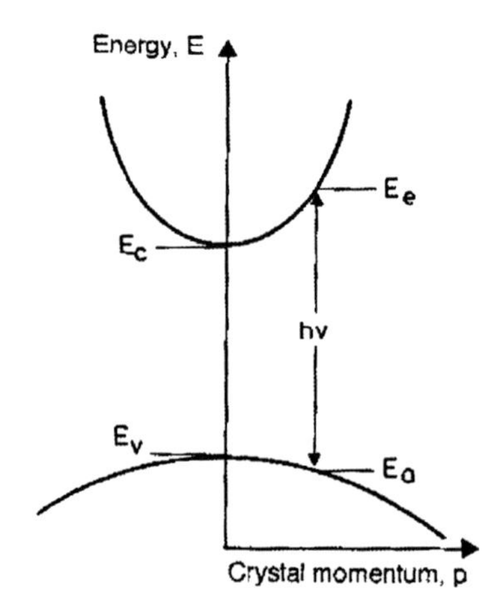

Fig. 2.4. Energy of conduction and valence band as a function of crystal momentum for a direct semiconductor. Also shown is absorption of a light quantum, hν

$p = \mathrm{h}\nu/c = \mathrm{h}/\lambda$, has a negligibly small momentum, p (h is the Plank's constant, ν is the frequency of the light, c is the velocity of light). The conservation principles of energy and momentum demand that during the absorption process the crystal energy rises, but the crystal momentum remains almost unchanged. This leads to certain selection rules.

The absorption process is best demonstrated in direct semiconductors. In Fig. 2.4 energy is plotted versus crystal momentum. In this representation, conduction band and valence band have a parabolic shape. In a direct semiconductor, the minimum energy of the conduction band in relation to the crystal momentum p lies directly above the maximum of the valence band. When a photon is absorbed, the energy $E = \mathrm{h}\nu$ is the energy difference between the initial and final condition of the energy of the crystal.

The situation is different in an indirect semiconductor. In this case, the minimum of the conduction band and the maximum of the valence band lie at different crystal momentums. It is, however, possible to excite it to the conduction band minimum if the necessary change in momentum can be induced by thermalvibrations in the lattice, i.e., a phonon. A phonon itself, although it only has a low energy level in comparison to a photon, has a very high momentum. The important point here is that the probability of absorption is much lower than for a direct semiconductor due to the involvement of two different particles.

Silicon is an indirect semiconductor, as depicted in Fig. 2.5. Therefore, it has a low absorption coefficient at photon energies near the band edge. This means that a relatively large thickness of material is necessary to absorb the long wavelength part of the solar spectrum.

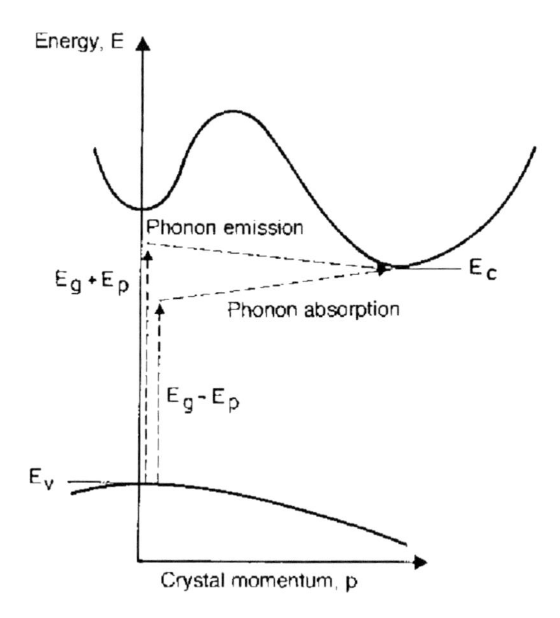

Fig. 2.5. Energy of conduction and valence band as a function of crystal momentum for an indirect semiconductor. Absorption at the gap energy is only possible by interaction with phonons

Recombination, Carrier Lifetime

If "excess" charge carriers are created in a semiconductor, either by the absorption of light or by other means, the thermal equilibrium is disturbed, then these excess charge carriers must be annihilated after the source has been "switched off". This process is called recombination.

The most important mechanisms for recombination are radiative recombination and recombination via defect levels.

Radiative recombination is when electrons "fall back" from the conduction band into the valence band, thus annihilating the same number of holes. The process is the exact inverse to absorption, and it is clear that this recombination energy must correspond to the energy Eg of the band gap. In silicon, this recombination is just as unlikely as absorption, which means that indirect semiconductors should have long charge carrier lifetimes. In silicon, the dominating recombination mechanism is via levels in the gap. It is a known fact that the lifetime in semiconductors is determined fundamentally by the presence of impurities and crystal defects. It is plausible that the inclusion of atoms that do not have the electron structure of a pentavalent or trivalent dopant will give rise to defect levels, with energy levels that need not lie near the edge of the band. They may lie deeper in the forbidden band and are thus called deep defects. Figure 2.3 shows a number of these energy levels for different substances in silicon. These impurity levels, also called "trap levels" because they are traps for charge carriers, determine the recombination of charge carriers to a high degree. For an energy level in the forbidden band, four fundamental processes are possible:

- an electron is captured by an unoccupied energy level (1);
- an electron is emitted from an occupied level into the conduction band (2);
- a hole is captured by an occupied energy level (3);
- a hole is emitted into an unoccupied state in the valence band (4).

The closer an energy level is situated to the middle of the gap the higher its efficiency as a recombination center. In addition, its atomic properties are influential. They can be characterized by its capture cross sections for electrons and holes. The quality of a semiconductor material is expressed by the lifetime τ for minority carriers. Obviously, good material has a long lifetime. For solar cells, even more important is the diffusion length L, which is derived from lifetime. It is the distance an excess carrier can move by diffusion before being annihilated by recombination. Only light absorbed within a distance of about the diffusion length from the p-n junction can contribute to the electrical output. High efficiency cells must have a diffusion length larger than the cell thickness.

2.2 The Silicon Solar Cell

The physics of solar cells is most straightforward for crystalline silicon cells [13]. To understand the function of semiconductor devices and thus of solar cells, a precise understanding of the processes within a p-n junction is crucial. The base unit of many semiconductor devices is a semiconductor body, in which two different dopants directly adjoin one another. This is called a p-n junction if a p-doped area merges into an n-doped area within the same lattice.

In a simple example, we assume that – in silicon – both dopants are of the same magnitude and merge together abruptly. Figure 2.6 may clarify this behavior. The left-hand side $x < 0$ would, for example, be doped with boron atoms with a concentration of $\mathrm{N_A} = 10^{16}$ atoms per cm^{-3}, making it p-conductive. The right-hand side $x > 0$, on the other hand, could be doped with phosphorus atoms, at $\mathrm{N_D} = 10^{16}\,\mathrm{cm}^{-3}$, making it n-conductive.

The freely moving charge carriers will not follow the abrupt change in concentration from $\mathrm{N_A}$ to $\mathrm{N_D}$. Rather, the carriers will diffuse due to the difference in concentration, i.e., the holes from the p region will move into the n region, and the electrons from the n area will move into the p region. Diffusion currents will arise. The ionized acceptors and donors, which are no longer electrically compensated, remain behind as fixed space charges (Fig. 2.6). Negative space charges will arise on the left-hand side in the p region, and positive space charges arise on the right-hand side in the n region. Correspondingly – as occurs in a plate capacitor – an electric field is created

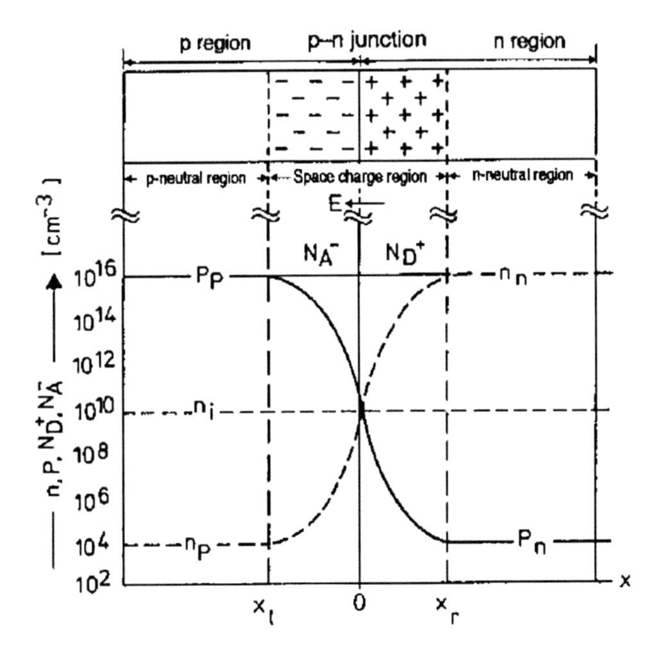

Fig. 2.6. Doping and concentration distribution of a symmetrical p-n junction in thermal equilibrium

at the p-n junction, which is directed so that it drives the diffusing charge carriers in the opposite direction to the diffusion. This process continues until an equilibrium is created or, in other words, until the diffusion flow is compensated by a field current of equal magnitude. An (extremely large) internal electric field exists – even if both sides of the semiconductor are grounded.

When the p-n junction is illuminated, charge carrier pairs will be generated wherever light is absorbed. The strong field at the junction pulls minority carriers across the junction and a current flow results. The semiconductor device is not in thermal equilibrium, which means that electric power can be delivered to a load. This is the basic mechanism of a solar cell. A typical such solar cell according to Fig. 1.1 consists of a p-n junction, which has a diode characteristic. This characteristic can be derived from standard solid state physics. It is:

$$I = I_0\big(\exp\left(V_\mathrm{A}/V_\mathrm{T}\right) - 1\big)\,, \qquad (2.2)$$

where I is the current through diode at applied voltage V_A. V_T is a constant, the so-called thermal voltage. I_0 is the diode saturation current, which depends on the type, doping density, and quality of the semiconductor material and the quality of the p-n junction.

If this junction is illuminated, an additional current, the light-generated current I_L is added:

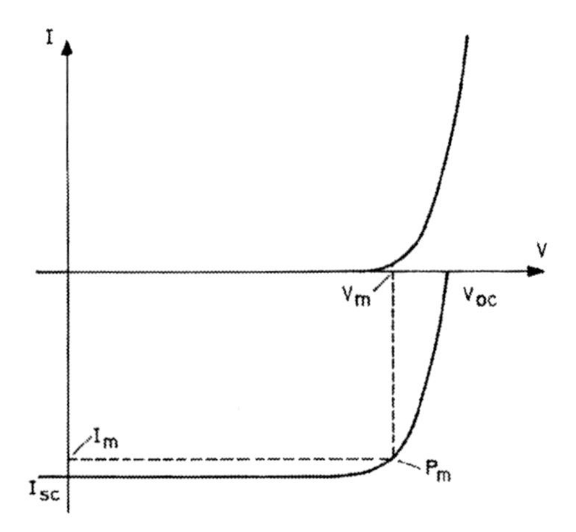

Fig. 2.7. I–V-characteristic of solar cell without (**top**) and with illumination. V_m, I_m, and P_m are values at maximum power

$$I = I_0\big(\exp(V_A/V_T) - 1\big) - I_L\,. \tag{2.3}$$

The negative sign in (2.3) results from polarity conventions. Now the current I is no longer zero at zero voltage but is shifted to I_L. Power can be delivered to an electric load. The I/V characteristic with and without illumination is shown in Fig. 2.7.

This figure also defines three important quantities: V_{oc}, the open circuit voltage, I_{sc}, the short circuit current, which is identical to I_L, and the maximum power point P_m at which the product of V and I is at a maximum. This is the optimal operating point of the solar cell. Voltage and current at P_m are V_m and I_m. It is obvious that the ideal solar cell has a characteristic that approaches a rectangle. The fill factor $FF = I_m V_m / I_{sc} V_{oc}$ should be close to one. For very good crystalline silicon solar cells, the fill factors are above 0.8 or 80%. From (2.3) we can also recognize the importance of the saturation current I_0. The open circuit voltage is obtained when no current is drawn from the cell. Then:

$$V_{oc} = V_T \ln\big(I_L / I_0 + 1\big)\,. \tag{2.4}$$

Even at low current densities the term I_L / I_0 is large compared to 1, so we find that $V_{oc} \approx V_T \ln(I_{sc}/I_0)$, i.e., the open circuit voltage is proportional to the logarithm of the ratio of I_{sc} to I_0. This means that although I_0 is a very small quantity compared to I_L, lowering the saturation current is very crucial for increasing efficiency. From solar cell physics it can be derived that there are three sources for I_0: a) minority carrier leakage current from the emitter region, b) a minority carrier leakage current from the base region, and c) a space charge recombination current.

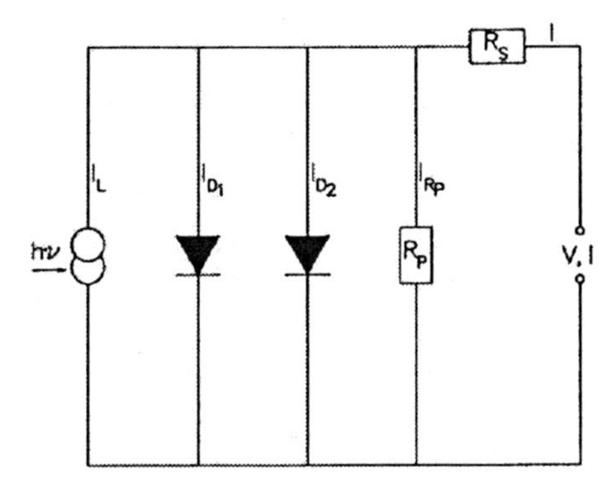

Fig. 2.8. Equivalent circuit of solar cell with two diode model

With these components an equivalent circuit of a solar cell can be constructed. It contains all relevant components. These are: a current source hν due to the light-induced current I_{L}, and two diode saturation currents I_{D1} and I_{D2}. The saturation current has to be represented by two diodes because the space charge recombination current has a different dependence on voltage than the other two currents. The other components are of resistive nature, a parallel (shunt) resistance R_{P} and a series resistance R_{S}. Evidently, R_{P} should be as high and R_{S} as low as possible.

Efficiency

The conversion efficiency is the most important property of a solar cell. It is defined as the ratio of the photovoltaically generated electric output of the cell to the radiative power falling on it:

$$\eta = \frac{I_{\mathrm{m}}\,V_{\mathrm{m}}}{P_{\mathrm{light}}} = \frac{FF\,I_{\mathrm{sc}}\,V_{\mathrm{oc}}}{P_{\mathrm{light}}} \tag{2.5}$$

where FF is the fill factor $V_{\mathrm{m}}I_{\mathrm{m}}/V_{\mathrm{oc}}I_{\mathrm{sc}}$ as further explained in Sect. 6.1.2. Efficiency is measured under standard conditions (see also Chap. 10).

3 Silicon Solar Cell Material and Technology

3.1 Silicon Material

Apart from oxygen, silicon is the most abundant element in the surface of the earth. It almost always occurs in oxidized form as silicon dioxide, as in quartz or sand. In the refining process, SiO_2 is heated to about 1800°C together with carbon. The metallurgic grade silicon that results from this process is used in large quantities in the iron and aluminum industries. Since it is only about 98% pure, it is not suitable as a semiconductor material and has to be further refined. This is done by transferring it into trichlorosilane ($SiHCl_3$), which is a volatile liquid. This liquid is distilled and subsequently reduced by reacting it with a hot surface of silicon, the Siemens process. Those two processes require a considerable input of energy and are the major contribution to the energy content of silicon solar cells.

3.2 Monocrystalline and Multicrystalline Silicon

3.2.1 Technology of Czochralski and Float Zone Silicon

In the beginning, only Czochralski (Cz) grown single crystals were used for solar cells. This material still plays an important role. Figure 3.1 shows the principle of this growth technique. Polycrystalline material in the form of fragments obtained from highly purified polysilicon is placed in a quartz crucible that itself is located in a graphite crucible and melted under inert gases by induction heating. A seed crystal is immersed and slowly withdrawn under rotation. At each dipping of the seed crystal into the melt, dislocations are generated in the seed crystal even if it was dislocation free before. To obtain a dislocation-free state, a slim crystal neck of about 3 mm in diameter must be grown with a growth velocity of several millimeters per minute. The dislocation free state is rather stable, and large crystal diameters can be grown despite the high cooling strains in large crystals.

Today, crystals with diameters of 30 cm and more are grown routinely for the semiconductor market. For solar cells smaller diameter crystals are grown because the usual solar cell dimensions are 10 cm by 10 cm or sometimes 15 cm

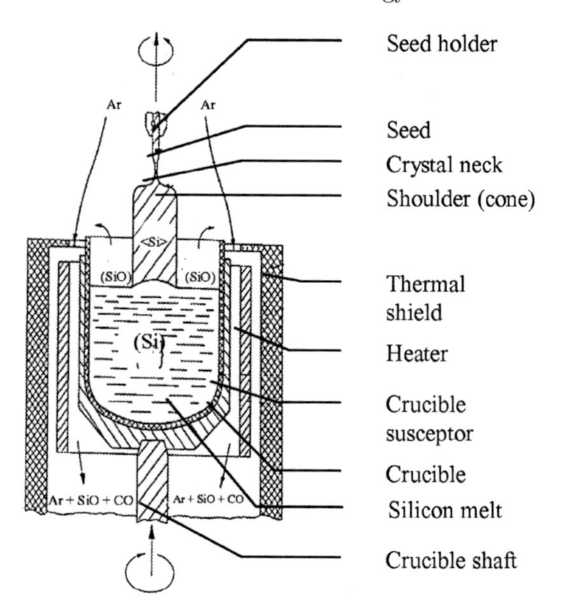

Fig. 3.1. Principle of the Czochralski growth technique

by 15 cm. The round crystals are usually shaped into squares with rounded corners in order to obtain a better usage of the module area.

The silicon melt reacts with every material to a large extent. Only silica can be used as a crucible material, because its product of reaction, silicon monoxide, evaporates easily from the melt. Nevertheless, Czochralski-grown crystals contain 10^{17}–10^{18} cm^{-3} of mainly interstitial oxygen. An alternative crystal growth technique is the float zone technique (Fig. 3.2). A rod of solid, highly purified but polycrystalline silicon is melted by induction heating and a single crystal is pulled from this molten zone. This material is of exceptional purity, because no crucible is needed, but it is more costly than Czochralski (Cz) material. In particular, it has a very low oxygen contamination which cannot be avoided with the Cz material because of the quartz crucible. Float zone (Fz) material is frequently used in R&D work. Record efficiency solar cells have been manufactured with float zone material, but it is too expensive for regular solar cell production, where cost is of overriding importance.

An interesting new development concerns tricrystals [14]. These are round crystals consisting of three single crystals arranged like pieces of a pie. They can be grown much faster and have higher mechanical stability. Solar cells of 0.1 mm thickness can be manufactured with a saving of 40% of the material.

For solar cells, as well as for all other devices, the crystal rods are separated into wafers of 0.2 mm to 0.5 mm thickness by sawing. This is a costly process because silicon is a very hard material that can only be cut with diamond-coated sawing blades. The standard process was the ID (inner diameter) saw, where diamond particles are imbedded around a hole in the saw

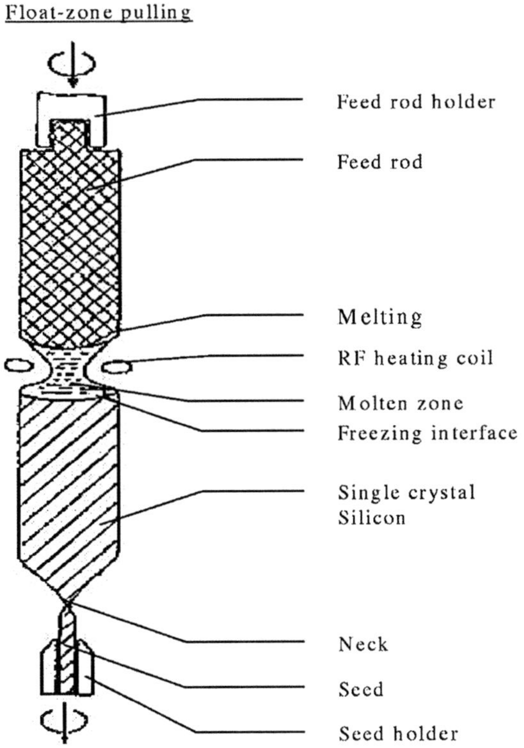

Fig. 3.2. Principle of the float zone technique

blade. A disadvantage of this process is that up to 50% of the material is lost in the sawing process. A new process was developed especially for solar cell wafers, the multi-wire saw (Fig. 3.3). A wire of several kilometers in length is moved across the crystal and wetted by an abrasive suspension whilst being wound from one coil to another. In this manner, thinner wafers can be produced and sawing losses are reduced by about 30%. It is interesting to note that wire saws are now also used for other silicon devices, an example of synergy in this field.

Another technology dating back to the 1970s is block casting [15], which avoids the costly pulling process. Silicon is melted and poured into a square graphite crucible (Fig. 3.4). Controlled cooling produces a polycrystalline silicon block with a large crystal grain structure. The grain size is some millimeters to centimeters and the silicon blocks are sawn into wafers by wire sawing, as previously mentioned. Cast silicon, also called polycrystal silicon, is only used for solar cells and not for any other semiconductor devices. It is cheaper than single crystal material, but yields solar cells with a somewhat lower efficiency. An advantage is that the blocks easily can be manufactured into square solar cells in contrast to pulled crystals, which are round. It is

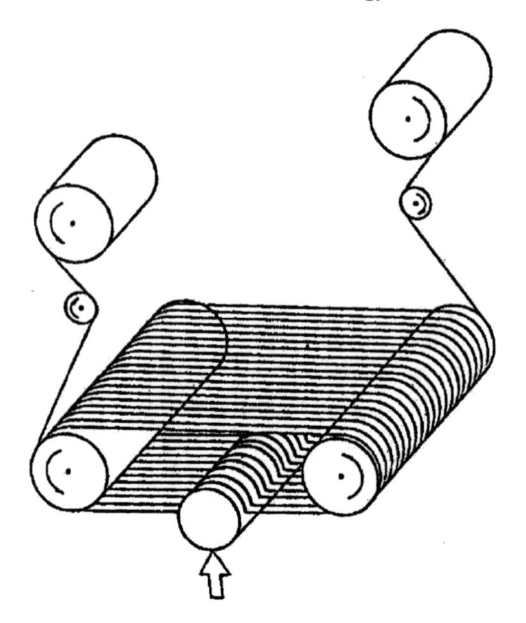

Fig. 3.3. Multi-wire sawing process

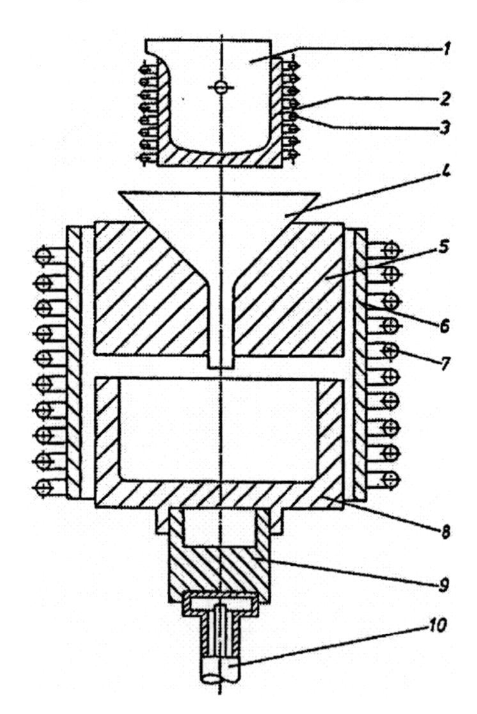

Fig. 3.4. Block casting apparatus

much easier to assemble multicrystalline wafers into modules with nearly complete utilization of the module area. Thus, the lower efficiency of cast material tends to disappear at the module level. Because of the contact with the crucible, polycrystalline silicon has a higher impurity content and thus lower carrier lifetime and lower efficiency than monocrystalline silicon. Point defects and grain boundaries act in the same direction. Several techniques have been devised to remove impurities during solar cell processing. Mobile impurities can be pulled to the surface by phosphorus gettering [16], which occurs during emitter diffusion. Immobile point defects are deactivated by hydrogen passivation. Atomic hydrogen can diffuse into silicon even at relatively low temperatures. Processed wafers are exposed to atomic hydrogen produced in a plasma discharge.

3.2.2 The Silicon Supply Problem

A big question mark for the future is related to the source of highly purified silicon for solar cells. Fifty percent of the cost of a module is due to the cost of processed silicon wafers. The PV industry has in the past used reject material from the semiconductor industry that was available at low cost. This created a dependence that is only viable if both sectors grow at the same rate. An additional problem is that the semiconductor market is characterized by violent cycles of boom and decline superimposed on a relatively steep growth curve. In boom times, the materials supply becomes tight and prices increase. This happened in 1998 when even reject material was in short supply and some solar cell manufacturers had to buy regular semiconductor-grade material at high cost.

One of the keys for cost reduction is to reduce the silicon content of the product. Present lines of approach are the reduction of kerf loss by wire sawing and the use of thinner wafers. The most advanced production lines use wafers of less than 0.2 mm thickness. Thinner wafers are also desirable, because if the right technology is used, efficiency is increased [17].

If the present standard technology is to continue its dominance, a dedicated solar-grade silicon will have to be developed. Even if only a 15% annual growth rate of the market is assumed, there will be a shortage of 5,000 megatons by 2010, which is two-thirds of demand [18]. Efforts to produce such material have been undertaken in the past but were not successful for two reasons: Purity requirements for solar silicon are very high, because photogenerated carriers have to be collected over large distances in such solar cells. This demands high carrier lifetimes and therefore an extremely low concentration of relevant impurities. This situation is aggravated by the continued trend toward higher efficiencies. The second point is that a dedicated solar-grade manufacture is only economical with large scale production. The present market would have to grow by about another factor of five in order to justify such manufacture. The Solar World AG recently announced plans to set up a large scale, 5,000 megatons production plant for solar silicon that could become operational within a few years. The process steps are [18]:

- trichlorosilane production from silicon tetrachloride, hydrogen chloride, and hydrogen;
- trichlorosilane pre-purification and silicon tetrachloride recycling;
- One step of trichlorosilane and silicotetrachloride recycling to silane and silicon tetrachloride redistribution;
- silane fine purification;
- thermolytic decomposition of silane to solar grade silicon granules in a fluid bed reactor.

3.3 Ribbon Silicon

3.3.1 Principle

Ribbons of silicon can be cast or grown by several techniques. The goal of crystalline ribbon technologies is to reduce cost by eliminating the costly silicon sawing process and at the same time minimizing the amount of silicon due to a reduced layer thickness and elimination of kerf loss. Supposing sufficient bulk quality, the resulting ribbons can be used directly as wafers for solar cell processing. If low-quality materials like metallurgical-grade silicon are used, a subsequent epitaxial growth of a highly pure active silicon layer is mandatory. In this case, the ribbons are used as a mechanical substrate and as an electrical conductor to the back electrode.

There have been numerous activities in the 1980s in the field of silicon ribbon growth for photovoltaic applications, which are described in [19–22] Out of the over twenty different approaches that had been under investigation, only two are commercialized. They will be described in the following section.

3.3.2 The Main Approaches in Ribbon Silicon Production

The Edge Defined Film Fed Growth Process (EFG)

In the EFG process developed by ASE Americas, a self-supporting silicon ribbon is pulled from the melt through a die which determines the shape of the ribbon (Fig. 3.5) [23, 24]. Today, octagon tubes of 5.3 m in length at a nominal average wall thickness of 280 μm are pulled out of a graphite crucible containing liquid silicon and are subsequently separated by a Nd:YAG laser [25]. The resulting sheets of $10 \times 10\,\mathrm{cm}^2$ have a somewhat lower material quality than single crystals, and they have a wavy surface. Nevertheless, conversion efficiencies of up to 14.8% were achieved in the production line with an excellent overall yield of over 90% at the moment and 95% expected in the

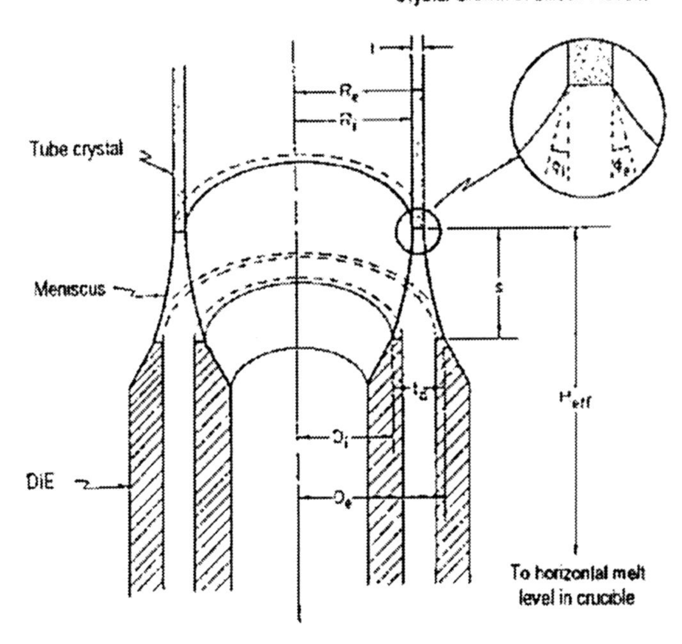

Fig. 3.5. Principle of EFG process

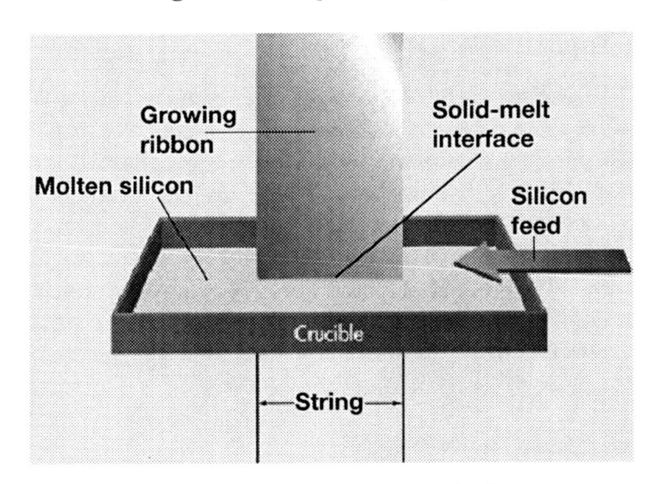

Fig. 3.6. The string ribbon process by Evergreen

near future. However, the long-term goal of this approach is to manufacture tubes of cylindrical shape with a diameter of 1.2 m and a wall thickness of about 100 μm (Fig. 3.6) [26]. In the meantime, ASE has expanded the annual capacity of the wafer production to over 13 MW and installed an automated 6.5 MW capacity pilot solar cell manufacturing facility [27].

Fig. 3.7. Fifty cm diameter EFG cylinder

The String Ribbon Process

The String Ribbon Process has been under development since 1994 at the Evergreen Solar Inc. The principle is shown in Fig. 3.7 [28–30]. In a fairly simple procedure, silicon ribbons of variable thickness are pulled with temperature resistant strings directly out of the melt and are cut subsequently into the desired length with diamond tools. The growth speed is up to 25 mm/min, resulting in ribbons with a thickness of below 100 μm and a conversion efficiency of 15.1% on a lab cell of $1\,\mathrm{cm}^2$ area [34, 35]. There are 30 and 60 W modules commercially available consisting of cells of $5.6 \times 15 \times 0.025\,\mathrm{cm}^3$ size, and the production capacity is about to be expanded significantly [33].

3.4 Silicon Cell Technology

Cell technology converts the silicon wafers into solar cells. As a rule, p-type silicon is used in the photovoltaic industry. For wafer processing, the following steps are important.

3.4.1 Production of pn and pp^+ Junctions

The active junction (emitter) is very close to the front surface. At the rear of the device a high p-doping is introduced to reduce contact resistance and surface recombination. These junctions are realized by thermal diffusion. In the diffusion process, an electrically heated tube furnace with a quartz tube is used. Diffusion temperatures vary between 800°C and 1200°C. All high-temperature steps require very clean conditions in order to avoid contaminants. Diffusion sources are phosphorus for the emitter junction and boron for the so-called back surface field, which repels carriers from the high recombination back contact. The doping elements are introduced as liquid or gaseous compounds, e.g., phosphine (PH) or phosphorus oxychloride ($POCl_3$) for n-doping, and boron bromide (BBr_3) for p-doping. In industry the back contact is often generated by sintering the aluminum contact. Since Al is a group III element, it will also generate p-type doping. This procedure, however, cannot be used for very high-efficiency solar cells.

3.4.2 Oxidation Process

Oxidation of the silicon surfaces is an important step in solar cell manufacture. It is carried out in quartz tube furnaces like the diffusion process. The oxidation atmosphere can either be dry oxygen or wet, which means oxygen plus water vapor. SiO_2 surface layers have several functions. They passivate the surface by reducing surface states that act as recombination centers. They act as diffusion barriers in selected locations, and they provide mechanical and chemical protection of the sensitive surface against further processing and module manufacturing. On air they also provide some antireflection property, but in encapsulated cells other materials with higher index of refraction have to be used. Excellent passivation properties can also be obtained with plasma-deposited silicon nitride.

3.4.3 Electrical Contacts

In the laboratory, electrical contacts are mainly made by vacuum evaporation, which is a very controllable technique. In industry, screen printing of thick films, which is more amenable to automation, is used. For the front contacts a paste consisting of 70% Ag, an organic binder, and sintered glass is applied. After deposition, the layer is sintered at approximately 600°C and a good electrical contact results. Screen printing has several disadvantages. Its lateral resolution is limited, and it only makes good contact to highly doped surfaces. For the back contact, a paste containing aluminum is also screen printed over the entire surface and then sintered. These and other simplifications explain why industrial solar cells have significantly lower efficiency but also lower cost than the best laboratory cells.

3.4.4 Antireflection Technologies

Two measures, which can be combined, are instrumental in reducing surface reflection. The first is surface texturing, which bends incoming rays into a

Fig. 3.8. Electron microscopic picture of silicon wafer with random pyramids

more horizontal direction and thus increases its path length inside the silicon (see Fig. 4.1). Surface texturing works only on monocrystalline silicon of ⟨100⟩ orientation. For texturing, the silicon wafers are immersed in a weak solution of KOH or NaOH at 70°C. In this way the ⟨111⟩ orientations are exposed as random pyramids (Fig. 3.8). Chemical texturing cannot be applied to cast multicrystalline silicon because it consists of crystal grains of different orientations.

The other possibility is to deposit a transparent antireflection layer. For this purpose, a material has to be used for which the refractive index is $n = (n_{\mathrm{si}} n_{\mathrm{s}})^{1/2}$, where n_{s} is the refractive index of the surrounding medium. Since in industrial production the cells are embedded into modules with a glass cover, its index has to be used. A practically applied material is titanium dioxide (TiO_2). It can also be incorporated into a screen printing paste and treated as described before. An antireflection film can also be combined with surface texturing for high efficiency cells.

3.4.5 Status Today

Today's photovoltaic market is characterized by the following trends:

- slow but steady improvement of conversion efficiency,
- slow reduction of the cost of modules and systems,
- uncertain supply base of polycrystalline raw material.

In this context, the importance of conversion efficiency has to be discussed. It could be argued that efficiency is not that important, provided the cells are very cheap, but reality has demonstrated that solar cells should have a minimum efficiency of about 10% in order to be useful (except for application in consumer products). This has to do with area-related cost, which constitutes a large part of systems cost. The solar cells have to be hermetically encapsulated in modules that are held in support structures and require electric wiring. All these factors depend on area and have strong influence on the cost of photovoltaic electricity. Therefore, both in research laboratories and in manufacturing, improvement of efficiency is a high priority. The best laboratory efficiency for single crystal silicon is 24.5% today [34]. This efficiency can only be realized with very elaborate technology. Experience has shown that progress in laboratory efficiency leads to improvement in production with a certain time delay. The best production cells now have an efficiency of 15 to 16% .

3.5 Advanced Si-Solar Cells

3.5.1 High Efficiency Cells

A real solar cell has a number of loss mechanisms, all of which can be minimized. The record efficiencies that have been achieved in the past have been accomplished mainly by careful attention to loss mechanisms. A survey of loss mechanisms can be seen in Fig. 3.9. Optical losses arise from reflection losses at the semiconductor surface. These can be reduced by antireflection

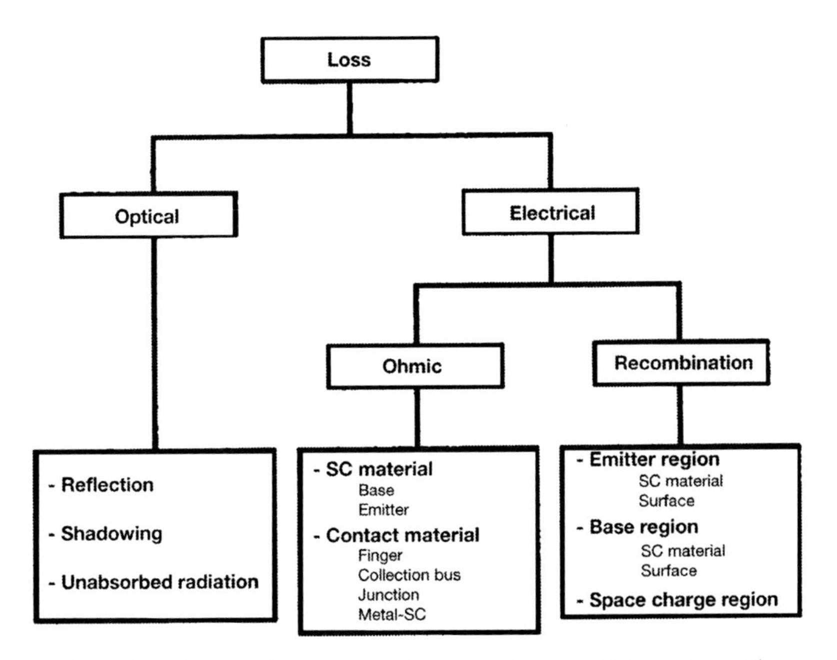

Fig. 3.9. Survey of loss mechanisms in solar cells (SC is solar cell)

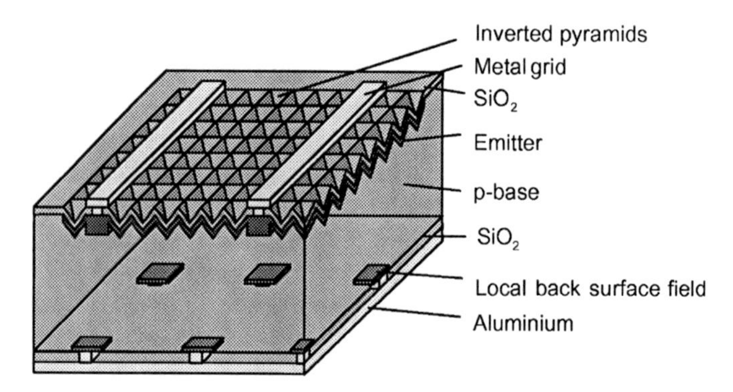

Fig. 3.10. Structure of a high efficiency monocrystalline solar cell

coatings and surface texturing. The electrical losses can be subdivided into ohmic and recombination losses. The ohmic losses arise in the semiconductor material, particularly in the thin emitter layer. In order to reduce these losses, the emitter is covered by a grid of metal fingers, which also contribute to losses. The junction between metal contacts and the semiconductor both at the front and rear can represent a contact resistance, particularly if the contact is to a region of lower doping. These losses can be reduced by locally restricted highly doped regions below the contacts. Recombination losses occur because photogenerated minority carriers can recombine before reaching the pn-junction and thus are lost for current flow. All three regions of a solar cell – the emitter layer, the base, and the space charge region between emitter and base – contribute to recombination and have to designed accordingly.

Optimal design principles are incorporated in modern high efficiency cells like the LBSF (local back surface field) [35] cell shown in Fig. 3.10. The following details are important for the very good efficiency of the cell, which is of the order of 23%–24%:

- 200-µm-thick base layer of float zone silicon.
- Textured front side with inverted pyramids and antireflection coating to reduce reflection.
- Narrow metal contacts from fingers to local highly doped emitter regions. In this way, recombination losses at the emitter surface are minimized. For good surface passivation, low doping at the surface is required. On the other hand, good ohmic contact can only be made to highly doped regions that also cause high surface recombination velocity. This leads to very restricted, locally doped contact regions.
- The base contacts at the lower side are also restricted for the same reason.
- Good surface passivation.
- The back surface contact has reflective properties and reflects light that penetrates to the back of the cell.

3.5.2 Bifacial Solar Cells

If the thickness of a solar cell is of the order of the diffusion length or smaller, then carriers generated throughout the entire base volume can be collected and the device is light sensitive on both sides. The only change that has to be made is that the rear of the cell has to be open to the light by applying grid fingers similar to the front side. The module also has to have a glass cover on both sides. Such cells are available on the market today. They still have a somewhat lower efficiency on the rear side, but this could be remedied by better design. Laboratory cells have exhibited efficiencies above 20% on both sides.

Bifacial modules can have interesting applications. They can collect light on both sides and in a number of such applications, have higher combined yield than one-sided modules. We will come back to bifacial modules in the systems part of this book.

3.5.3 Buried Contact Cells

Another type of high efficiency solar cell is the buried contact Si solar cell shown in Fig. 3.11. This structure was first realized in 1985 by M. Green [36] and is patented in many countries. The significant difference in this cell is the buried contact. Using laser technology, grooves of approximately 20 μm wide and up to 100 μm deep to hold the grid fingers are cut in Si wafers textured according to the principle of random pyramids. The etching process that follows removes the silicon destroyed by this process. These grooves provide two advantages. Firstly, shadowing is reduced significantly when compared with the normal grid structure of commercial solar cells. Values of only 3% surface shadowing are obtained. Secondly, the grooves can be filled with contact metal. Thus, approximately the height of the grooves and thus the metallization can be five times its width. In conventional cells, even for vacuum-evaporated contacts and contacts reinforced by electrodepositing, the ratio

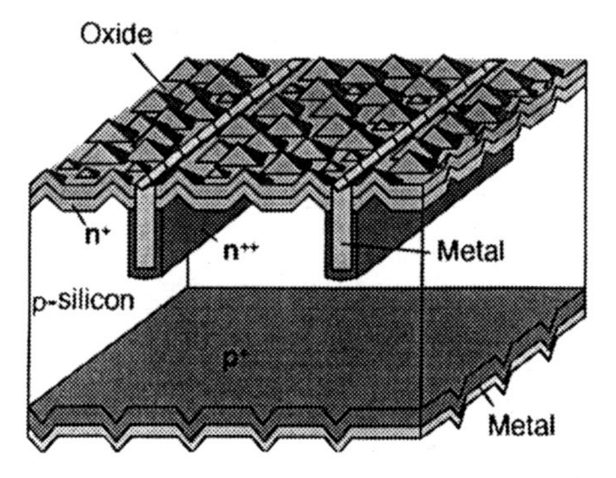

Fig. 3.11. The buried contact solar cell

is 1 : 1, i.e., the metal contact height is equal to its width. The five-fold height means a reduction in the resistance of the contact finger by a factor of 5, resulting in a significantly better fill factor. The technology of metalizing consists of an electroless-deposited nickel contact, which is sintered and then reinforced with copper. Since this technique does not require photomask processes or high-vacuum evaporation technologies, and is thus significantly more economic, it is well suited for use in serial production. The double-stage emitter is used for the emitter structure, whereby the highly doped n^{++} film is restricted to the grooves. The p^+ back surface field permits higher efficiencies.

A further advantage of this cell is the textured back surface, which increases the confinement of light and thus the total efficiency. With this type of cell (large area), average efficiencies of 18% have been achieved in production. With specific techniques, such as an improved antireflection coating and a local back surface field, efficiencies of up to 21% have been achieved in the laboratory [37].

3.5.4 Interdigitated Back Contact Cells

The IBC (Interdigitated Back Contact) structure suggested by Lammert et al. [38] in 1974 was realized by Swanson et al. [39] and Sinton et al. [40]. As the name suggests, both n^+ and p^+ contacts (Fig. 3.12) lie on the back of the cell. This structure has the following advantages.

- There is no shadowing of light by the finger structure.
- The metal contacts can be broad and take up almost the entire back surface. They are therefore of very low impedance, thus achieving a very low series resistance. Therefore, these cells are preferably applied in systems with light concentration.
- The penetration depth of n^{+-} and p^{+-} layers are noncritical.

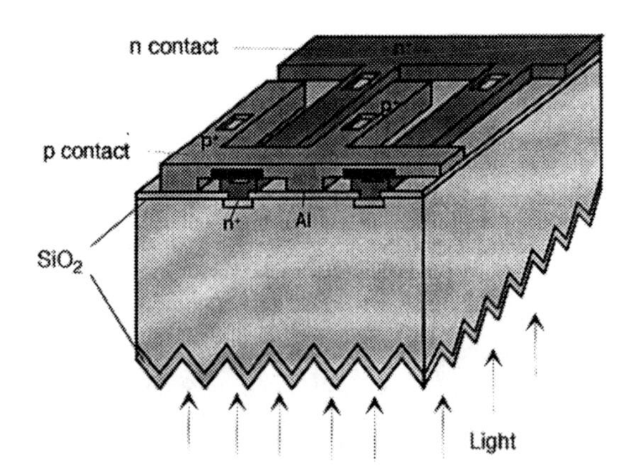

Fig. 3.12. The buried contact solar cell

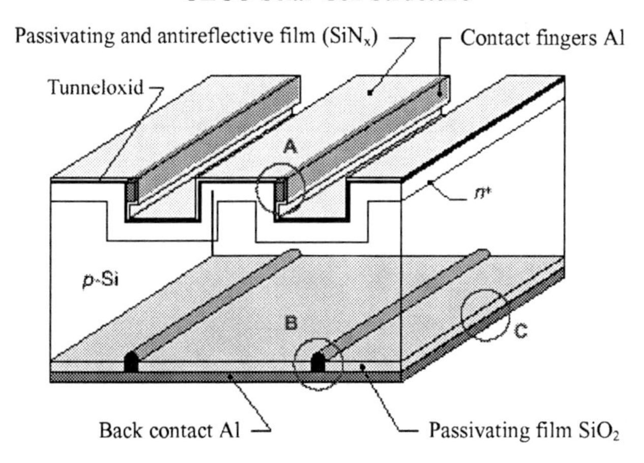

Fig. 3.13. The OECO (Obliquely Evaporated Contact) cell

This concept was further developed into the point contact solar cell in which only localized contacts in small points are made to the base layer. It requires very high quality silicon material and very good surface passivation. So far, it was only made for special applications where efficiency was of more importance than cost, but a simplified version of this cell is now being commercialized by the Sun Power Corporation for the general market.

3.5.5 OECO Cell

The OECO (Obliquely Evaporated Contacts) cell was developed by R. Hezel at the ISFH in Hameln [41]. This is a high efficiency cell that was specifically designed for easy manufacturability. Its basic features are shown in Fig. 3.13. The cell surface has mechanically machined grooves. The Al contacts are at the sides of the grooves such as to avoid light losses by shadowing. Evaporation of contacts is done by arranging large numbers of cells surrounding the evaporation source.

The latest results of this cell are: efficiencies of 20.0% for a 96 cm^2 cell with float zone silicon and 19.4% with czochralski silicon. This type of cell has also been realized as a bifacial cell.

3.5.6 a-Si/c-Si Heterostructures

A very interesting new development is the combination of crystalline and amorphous technologies in heterostructures. Absorption of sunlight still occurs mainly in a wafer of mono- or polycrystalline silicon. The silicon wafer is contacted on both sides with amorphous silicon films. The principle is shown in Fig. 3.14. This technique represents a combination of c-Si and a-Si technologies. This configuration has the following advantages:

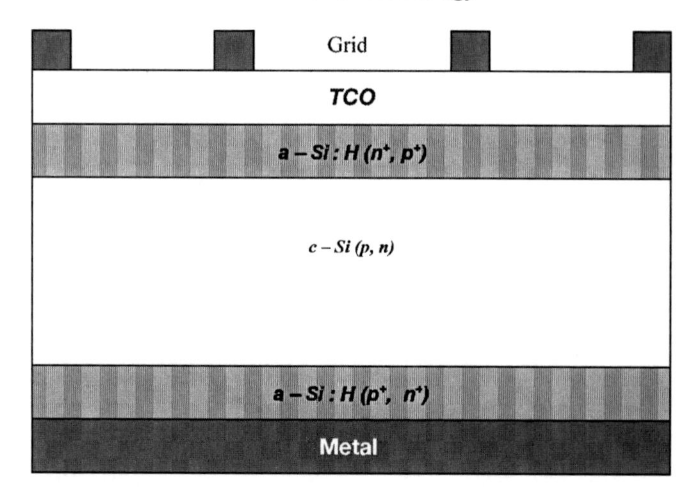

Fig. 3.14. Cross section of a Si/C-Si/a-Si heterostructure

- Potential for high efficiency.
- Very good surface passivation → low surface recombination velocity.
- Low processing temperatures → all processing steps occur below 200°C.
- Low thermal budget for processing → reduction of energy payback time.
- Reduced cost of cell technology.

The best results with this approach were obtained by the Japanese company Sanyo. The latest achievement is a conversion efficiency of 20.7% for a cell area of 101 cm^2 using n-type CZ-Silicon as the base (light absorbing) part [42].

Sanyo has developed the HIT-structure (Heterojunction with Intrinsic Thin layer). Its composition is a-Si(p^+)/a-Si(i)/c-Si(n)/a-Si(i)/a-Si(n^+). Very crucial in this structure is a thin intrinsic a-Si layer contacting the c-Si on both surfaces. This is apparently important for achieving good surface passivation. Interesting is that the crystalline silicon is n-type material. Because of the low conductivity of a-Si, a front window layer of a transparent conductor is required.

According to the information provided by Sanyo, the solar cells have excellent stability. Sanyo is now marketing this new type of cell.

3.5.7 Rear Side Contacted Cells

Recent developments in silicon solar cell technology exploit the advantages of thin wafers. Thinner wafers have economic and physical advantages. The most important effect of using thinner wafers is that the amount of expensive silicon material per cell is reduced. From the viewpoint of device physics, it can be shown that thin wafers can lead to higher efficiency, provided recombination at the back contact can be minimized.

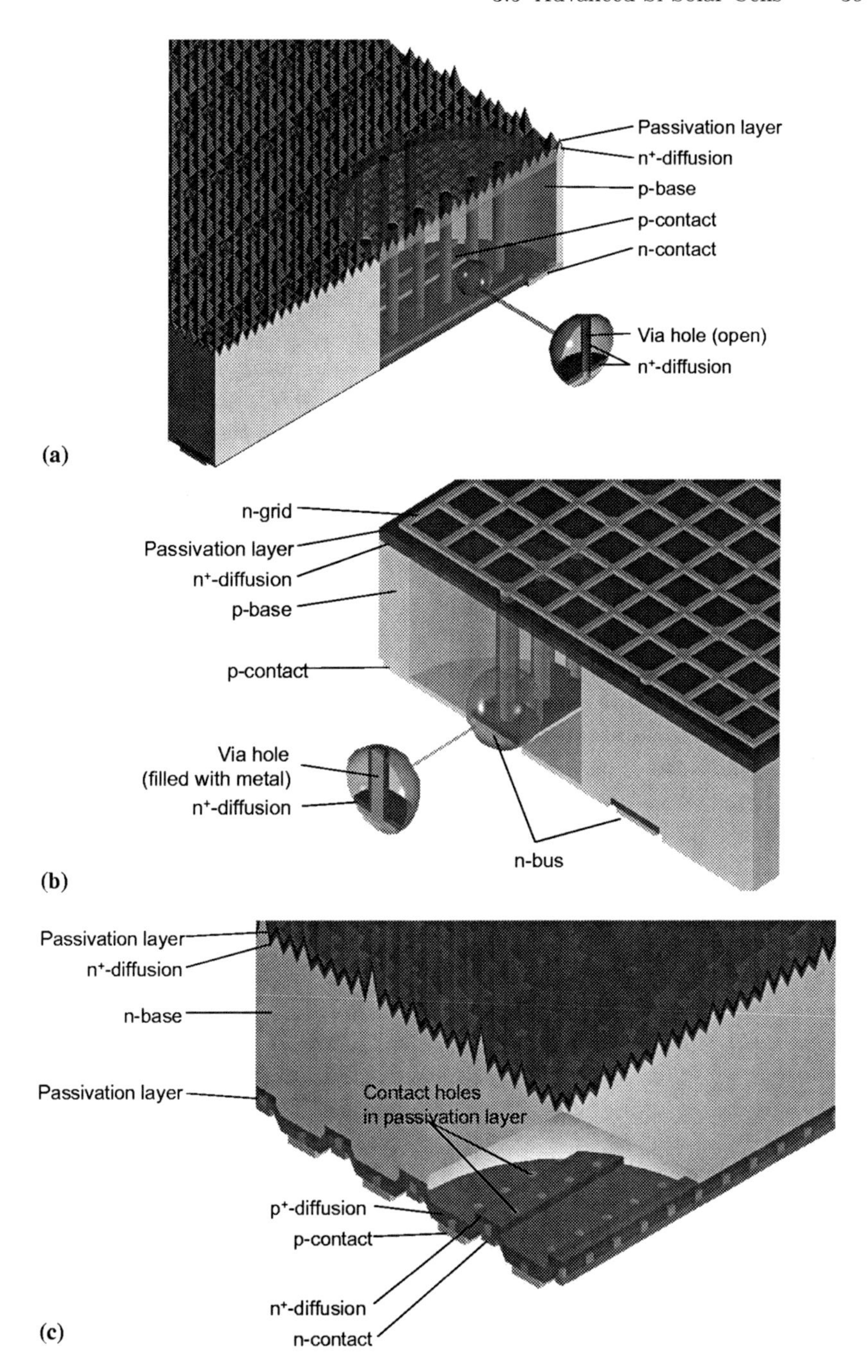

Fig. 3.15. (**a**) The emitter wrap through cell, (**b**) the metal wrap through cell, (**c**) the point contact solar cell

Thin silicon wafers permit new technologies for the contacts. Both n and p-type contacts can now be moved to the rear side. There are three different technological solutions for this task (Fig. 3.15a–c):

- the point contact cell which is an improved version of the interdigitated back contact cell shown in Fig. 3.12,
- emitter wrap through cell [43],
- emitter contact wrap through cell [44].

The last two types are made by laser drilling holes through the silicon. In the emitter wrap through cell, the n-type emitter is brought to the rear side of the cell and both emitter and base contacts are made there in interdigitated manner. The emitter diffusion also covers most of the rear of the cell. The emitter metal contact cell is similar, with the difference that the emitter metal contact fingers are still at the front side but connected through metal-filled holes to the rear. The difference of this technique is that fewer holes are required.

Such cells have the following advantages:

- Little or no shadowing of the front by metal grids, therefore, more light current.
- The front of the cells and modules has a very uniform appearance, an important aspect for architectural applications.
- Series connection of the cells in the modules is much simpler, because all connections are in the same plane.
- Lower quality silicon material can be employed, because the emitter junction is on both sides of the cell and, therefore, carriers have to diffuse not more than half the cell thickness.

3.5.8 Laser-Fired Contact Cells

Laser-Fired Contact (LFC) cells (Fig. 3.16) offer the possibility for mass production of high efficiency solar cells like the LBSF with local back surface field (see Fig. 3.10). The localized back contacts involve expensive photolithography steps and so are not attractive for production. The new technology avoids this step by rapid laser firing of the contacts. The rear side of the cell is first

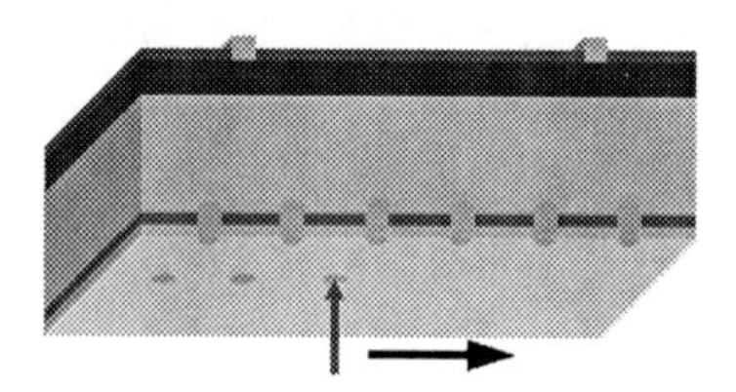

Fig. 3.16. The laser (*vertical arrow*) is scanned across the rear side of the cell, forming localized contacts

passivated by a silicon nitride film and then an aluminum layer is applied. Then, the local contacts are produced by scanning a q switched Nd-Yag laser across the wafer. An entire 10 cm by 10 cm wafer can be processed in about 2 sec. The contacts are at 1 mm centers. The laboratory results are excellent: Efficiencies up to 21.3% have been reached [45].

Advantages of LFC cells:

- No warping of wafers compared to complete Al-covered rear sides, very important for thin wafer processing.
- Very good light reflection at the rear side.
- The laser-fired contacts form functional back contact fields.

Very high efficiencies comparable to much more complex cells have been achieved.

4 Crystalline Thin-Film Silicon[1]

In view of what was said about the properties of silicon in the basic physics part, it would appear to be a contradiction to make a thin-film cell from crystalline silicon. Nevertheless, it has been shown both theoretically and practically that crystalline Si cells with a thickness of only a few micrometers can have reasonable efficiencies. The key is clever optical design of the film. By total internal reflection, light can be scattered and confined within a film, resulting in a path length exceeding film thickness thirty times or more. At the same time, it has to be ascertained that light-generated carriers do not recombine at the surfaces which are very close to the active areas. The term optical and electrical confinement has been coined to describe the operating principle of crystalline thin-film solar cells.

4.1 History

The first considerations concerning thinner silicon wafers for solar cells were made by M. Wolf and J. Lofersky while simulating the ideal parameters for record high efficiencies [47, 48]. They pointed out that with decreasing cell thickness the open circuit voltage increases due to the reduced saturation current, which again is a result of a decreasing active volume. In these papers, the benefits of a thickness on the order of 30 μm were quantified, also demonstrating the importance of low surface recombination velocities and a good optical confinement to make use of this advantage. The first theoretical work about light trapping in silicon layers was performed by Goetzberger, who suggested a Lambertian back reflector as a simple but efficient structure [49]. Different possibilities of light trapping are shown in Fig. 4.1.

The wide range of technological advantages incorporating the potential for a significant cost reduction was also described by Goetzberger et al. [50]. After these fundamental works, it took more than a decade until an increasing number of groups singled out the various technological problems and realized the first test solar cells to overcome these problems. Nowadays, the driving force for the development of crystalline Silicon Thin-Film Solar Cells

[1] This chapter as well as the succeeding one on thin-film materials follow in part a recent review by A. Goetzberger, C. Hebling, and H.-W. Schock [46].

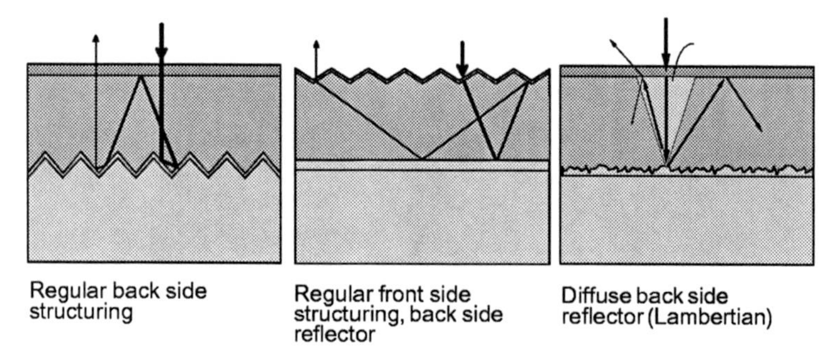

Fig. 4.1. Optical confinement for light path enhancement

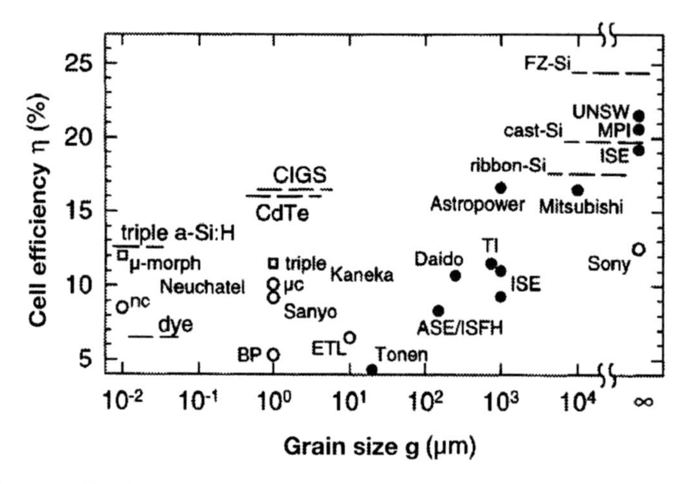

Fig. 4.2. Realized solar cell efficiencies as a function of grain size [46]

(c-SiTFC) is the inherent possibility for cost reduction, although this advantage has not yet been translated into commercial products.

An illustration of the conversion efficiencies as a function of the grain size achieved by the different institutes is given in Fig. 4.2.

4.2 The Basic Components of a Crystalline Silicon Thin-Film Solar Cell

The linking feature of all c-SiTFC approaches is the underlying substrate needed as a mechanical support due to the reduced thickness of the active silicon layer of typically 5 to 50 μm. The substrate consists either of low quality silicon like the previously mentioned SSP-ribbons or of foreign substrates such as glass, ceramics, or graphite. The choice of the substrate material determines the maximum allowed temperature for solar cell processing and,

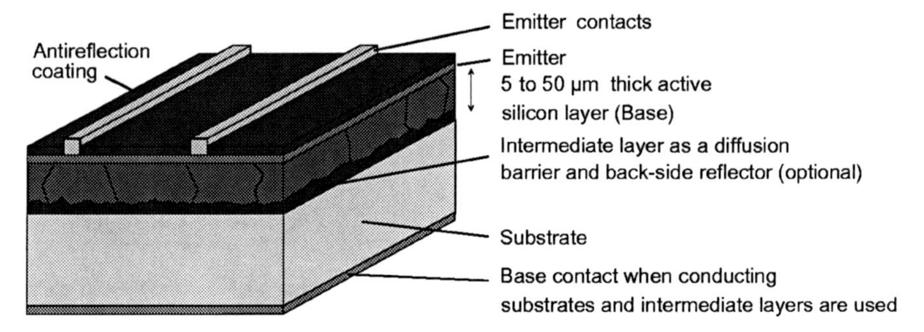

Fig. 4.3. The basic structure of a high-temperature crystalline silicon thin-film solar cell

therefore, nearly all c-SiTFC approaches can be assigned to one of three categories: (i) high temperature approach; (ii) low temperature approach; and (iii) transfer techniques, which are described below. Furthermore, the use of electrically conductive substrate materials enables the conventional front and back side cell; insulating materials allow for the monolithic series-interconnection of several cells. The most critical requirements concerning the substrate are low cost, thermal stability, a matching thermal expansion coefficient, mechanical stability, and a certain surface flatness [51, 52]. The basic structure of a thin-film cell of the type considered here is shown in Fig. 4.3.

Due to the high temperatures that arise in cell processing, impurities might migrate from the substrate into the active silicon layer with harmful effects on the conversion efficiency. Therefore, the deposition of barrier layers like SiO_2 or SiN_x on the substrate prior to the deposition of the electrically active silicon layer is an effective measure to suppress the impurity diffusion. Furthermore, such intermediate layers can act as a back-side reflector in order to achieve a good optical confinement.

There is a large variety of silicon deposition technologies that can roughly be allocated to the main groups of liquid phase and gaseous phase deposition. In the liquid phase deposition, the respective substrate is brought into contact with a metal melt (Cu, Al, Sn, In) saturated with silicon. By lowering the temperature of the melt, supersaturation occurs and silicon is deposited on the substrate. The substrate temperature lies within the range of 800–1,000°C, and deposition rates vary from a few μm/h to up to 10 μm/h. The term Liquid Phase Epitaxy (LPE) process is also used when the growth is not epitaxial, i.e., reproducing the crystal orientation of the substrate. In the Chemical Vapor Deposition (CVD) method, which is a well established method in microelectronics, a mixture of H_2 and the precursors SiH_4, SiH_2Cl_2, or $SiHCl_3$ is decomposed thermally at the hot surface of the substrate. The most common techniques are low pressure and atmospheric pressure CVD (LP-, AP-CVD), but also plasma-enhanced, ion-assisted, or hot-wire CVD (PE-, IA-, or HW-CVD) are used to deposit silicon films at temperatures between 300°C and

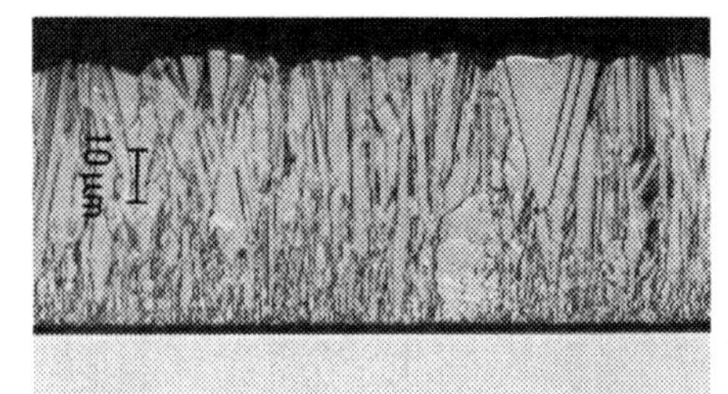

Fig. 4.4. Si layer deposited on a SiO_2 intermediate layer (**left**) and after a zone melting recrystallization step (**right**)

1, 200°C. There is little known about the economical aspects of the different deposition methods when used in large-scale production of solar cells, but there is a trend toward APCVD due to the potential for continuous inline processing and realized deposition rates of more than 5 µm/min.

As a general trend, the cell efficiency increases with the grain size. This is due to the fact that grain boundaries in their unpassivated state can be very effective recombination areas that reduce the diffusion length of the minority carriers drastically. On the other hand, it is well-known from simulations that the diffusion length should exceed the thickness of the active Si layer by at least two times in order to achieve high conversion efficiencies. Therefore, whenever the Si films are deposited on intermediate layers or foreign substrates, and thus are too fine-grained for acceptable minority carrier mobilities, an upgrading by recrystallization is recommended. Various methods have been used, depending on the procedure used to couple the energy into the Si layer and depending on the thermal budget that is allowed from the substrate. The most common recrystallization mechanisms that occur via the liquid phase are laser, electron-beam, strip heater or halogen lamp recrystallization. They are distinguished according to the form and size of the liquid zone, the pulling speed of the Si grains, the melting depth the three-dimensional temperature gradient and thus the grain size and defect density in the resulting Si layer, and finally the scalability of the apparatus (Fig. 4.4).

Finally, the solar cell technology for Si thin-film solar cells has to be adopted in several ways:

- Due to the weak absorbance of crystalline Si, the light trapping is one of the crucial measures that have to be applied in order to realize a high degree of total internal reflection. This can be achieved by a Lambertian back side reflector or by a textured front surface in combination with a reflecting back side.
- The electrical passivation of the surfaces is important in order to enhance the diffusion length of the charge carriers. This is done for example by an SiO_2 layer on the front side and by a thin highly doped Si layer forming a back surface field that prevents the charge carriers from recombination at the back side.

- If nonconducting intermediate layers or substrates are used, both the emitter and the base contacts have to be applied on the front side. The emitter contacting can be done conventionally, but the base must be contacted either by a selective emitter (the emitter is interrupted to make room for base contacts) or by trenches etched through the homogeneous emitter. Such a front contacting scheme enables the possibility of an integrated series-interconnection of several cells resulting in a monolithic solar cell module on a single substrate.
- For the solar cell formation, the layer system is processed with various temperatures due to the silicon deposition, silicon recrystallization, emitter formation, impurity gettering, electrical passivation of the surface or of the bulk silicon, deposition of antireflection layers or the deposition of the electrodes. The individual thermal steps as well as the total thermal budget are other crucial parameters for a successful manufacturing of c-Si thin-film solar cells.

4.3 The Present Status of the Crystalline Silicon Thin-Film Solar Cell

There is a broad range of research activities worldwide that has demonstrated the efficiency potential of thin crystalline silicon with conversion efficiencies of up to 21% under ideal conditions. Mostly, the aim is to prove the respective concept and to study the influence of the most crucial electrical and optical parameters. Generally, it must be noted that most realized cells were still made on Si wafers as substrates. Such test structures were made under ideal conditions using the best understood materials available, aiming to prove the concept and to study the influence of the different boundary conditions on the cell performance. The insights gained from these ideal systems are now about to be transferred into cost-effective layer-systems and processing technologies. The crystalline Si thin-film solar cell has seen a tremendous development in the last decade, although only Astropower transferred this effort into a commercially available product.

The large number of different approaches can roughly be classified into the three categories: i) Si layers deposited directly onto glass; ii) Si layers on high-temperature resistant substrates; and iii) semi-processed monocrystalline Si layers from Si wafers transferred onto glass. Some outstanding results obtained within this classification will be discussed in the following sections.

4.3.1 Si Layers Deposited Directly onto Glass

The inherent limitation of all approaches compatible with glass substrates is the melting point of glass, which is on the order of 600°C. This means that both the Si deposition and the solar cell processing is not allowed to

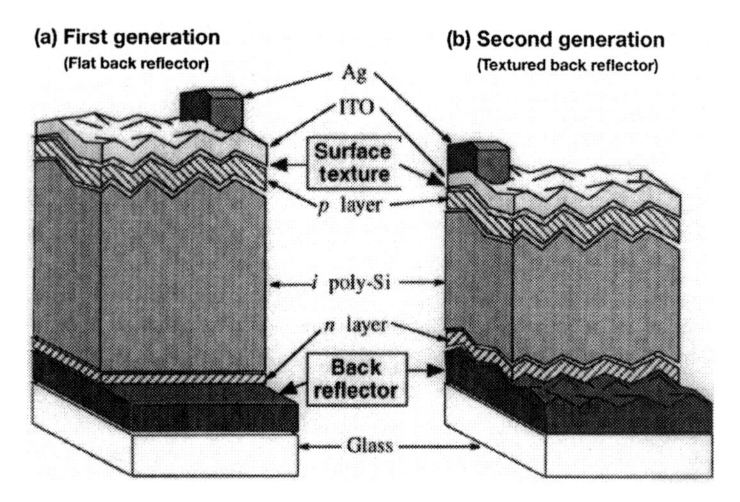

Fig. 4.5. STAR structure developed at Kaneka Corp., Japan

go beyond this temperature limit for a longer period of time. The crystallization of the Si layer can be done either by laser- or by solid-phase crystallization. Kaneka Corp., for example, utilizes a laser crystallization of a highly doped contact layer followed by the deposition of an intrinsic microcrystalline absorber film. The crucial feature of this "STAR" structure (Surface Texture and Enhanced Absorption with a Back Reflector) is a Si-layer thickness of less than 5 μm combined with an excellent optical confinement due to a rough reflecting back layer (Fig. 4.5) [53]. A triple cell stack of a a-Si:H/microcrystalline-Si/microcrystalline Si achieved a stable efficiency of 11.5%.

The Center for Photovoltaic Devices at the University of New South Wales and the Pacific Solar Corp. are exploring the so-called multijunction solar cell concept, applying the deposition of a high number of layers with alternating p- and n-doping levels. The basic idea is that very short minority carrier diffusion lengths can be tolerated in these layers and thus very impure material can be used. The conversion of this benefit into a good cell performance depends strongly on the electrical connection of the layers with the same polarity and on the recombination in the space charge region. Efficiencies of up to 17.6% were obtained with a six-layer system deposited epitaxially on a highly doped Cz wafer in order to proof the concept, but no results have been reported on a low cost system [54].

For its product development, however, Pacific Solar selected a simple one-layer system silicon on glass [55]. It consists of a polycrystalline Si film deposited on textured glass. The film is only less than 2 μ thick. This film is manufactured by a rapid low temperature deposition and a subsequent higher temperature treatment for improved electronic properties. Details of the process are not known. The structure of the film and module construction are

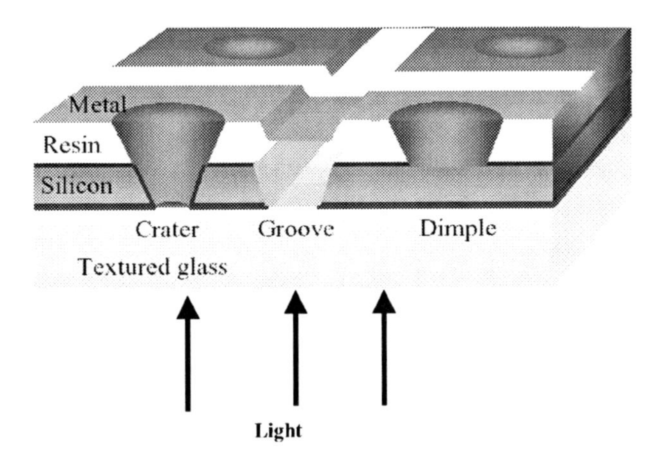

Fig. 4.6. Thin-film cell on glass by Pacific Solar

shown in Fig. 4.6. The thin n+pp+ silicon film is deposited on a glass superstrate, which means that the light is coming from below in the figure. The silicon is first covered by an insulating resin followed by the metal contact structure. P- and n-type contacts are made by an ingenious scheme shown in Fig. 4.6. Deep craters and shallower dimples are created by laser firing and connected to the metal leads. Single cells are separated by laser grooving and the cells are series-connected by the metal leads. In this manner monolithic series-connected modules are manufactured similar to thin-film modules from other materials, which will be described in the following chapter.

The pacific solar films have the interesting property that the film itself exhibits good electrical conductivity, which makes conducting oxides on the surface superfluous.

The best modules at present have an efficiency of 7.25%; average pilot production efficiency is 6.5%. An efficiency of 8% is expected for the near future and 10% within two years. The excellent durability of the modules has also been established. Current manufacturing cost is US $1.95/W with the expectation that with larger volume this will go down to US $1/W for larger production runs. This product is now in pilot production. Further development depends on finding the capital investment necessary for large-scale production.

4.3.2 Si Layers on High-Temperature Resistant Substrates

The high-temperature approaches rely on the high Si-deposition rates ($\cong 5\,\mu\text{m/min}$) at temperatures beyond 1000°C and Si-recrystallization techniques via the liquid phase ($T_{\text{melt,Si}} \sim 1420°\text{C}$). With these technologies, a high throughput, a continuous inline processing, large grains and, thus, high efficiencies can be obtained. However, a substrate material with ideal technical properties at a low price cannot yet be found. As a general rule, inexpensive materials usually contain a high level of impurities at high temperatures that causes the migration of these impurities from the substrates into the active Si layer. Furthermore, the complexity of the process steps for the layer preparation and the solar cell processing, particularly on insulating substrates, counts toward the critical side.

The Fraunhofer Institute for Solar Energy Systems (ISE) is investigating three different high-temperature approaches:

- Highly doped silicon ribbon pre-sheets are used as substrates for a subsequent direct epitaxial growth of the active Si layer in a rapid thermal chemical vapor deposition (RT-CVD) reactor [56]. The thermal processing of the layer system for solar cells has to take into account that the impurities migrate into the active Si layer and thus the maximum allowed impurity level in the substrate has to be determined.
- A highly doped Si sheet of low quality as a substrate, which is covered subsequently by a SiO_2 layer with via-holes. The total area coverage of the holes is 0.5%, which is enough to maintain a sufficient electrical contact to the back electrode, but still suppresses the diffusion of impurities from the substrate significantly. The top of the CVD-deposited Si layer is exposed to a large area recrystallization step in order to enlarge the grain size, and the seeding occurs via the Si crystals that are located underneath the via-holes.
- Encapsulated foreign substrates such as ceramics or graphite are used and the upgrading of the fine-grained Si layer occurs with a zone melting recrystallization step. An interdigitated front contact scheme was developed for such SOI (Silicon-On-Insulator) structures that allows the emitter and the base electrode formation at the front side. An efficiency of 11% was realized on an SiC-covered graphite substrate from ASE [57], as well as 9.4% on silicon nitride [58]. In order to study the potential of such a front-contacted SOI structure (silicon on insulator), epitaxially grown CVD layers on SIMOX-wafers were processed to solar cells, which resulted in efficiencies of up to 19.2% (Fig. 4.2). Additionally, 24 series-connected cells of 1 cm^2 size each on one SIMOX wafer using such an interdigitated front grid achieved an open circuit voltage of 15.2 V.

The approach of Mitsubishi Corp. also relies on their knowledge of the SOI technology and focuses on the formation of a high quality mc-Si layer on an insulator. A 3-μm-thick LPCVD layer deposited on a SiO_2 layer is zone-melted in a strip heater, resulting in grains of cm-length and defect densities on the order of some 10^6 cm^{-2}. After the epitaxial growth of the 60-μm-thick active Si layer, the supporting Si wafer is etched back by a patterned mask for back electrode formation. The 16.45% conversion efficiency is the highest value obtained by means of a recrystallized Si layer [58]

The only thin Si cell concept that made it to a commercially available product is the Silicon Film™ technology developed by Astropower [59]. Very little is known about this approach, but the active Si layer is probably deposited on a micro-grooved, conducting ceramic substrate that is covered by an intermediate layer as a back reflector and diffusion barrier. The grains are about 2 mm in diameter, and the electrical contact to the back electrode is made by via-holes in the dielectric layer. As important process steps, impurity gettering and hydrogen passivation serve to boost the minority carrier diffusion length. A remarkable efficiency of 16.6% was achieved on 1 cm^2 in the laboratory; cells of 15 cm in width and variable length went into production.

4.3.3 Transfer Technologies of Monocrystalline Thin Si Films onto Glass

The approaches based on transfer technologies of thin monocrystalline Si films are the latest and very exciting developments. The basic idea is to detach a semi-processed thin monocrystalline Si layer from the monocrystalline wafer and transfer it onto a substrate, preferably glass. The advantages are obvious: The superior material quality enables very high efficiencies even from less than 40-μm-thick Si layers; inexpensive glass can be used as substrate; and finally, the wafer can be reused for the thin-film formation and detachment several times.

The basic process sequence is shown in Fig. 4.7. This sequence shows the most practical approach: the layer transfer with a porous silicon separation layer. Other means of separating a thin surface layer also exist but present more technical difficulties.

The first real Si-transfer approach was pursued by R. Brendel with the so-called Ψ process. A textured [100]-oriented Si wafer with a porous Si film serves as an epitaxial seed for the growth of an approximately 10-μm-thick monocrystalline Si film. The epitaxial growth serves to conformally coat the surface of the anodically etched porous Si, resulting in a waffle-like structure enclosed by (111) planes [60]. Simulations predicted efficiencies of up to 19% under ideal technological conditions.

The Epi-lift process, which was developed at the Australian National University (ANU), is based on the formation of (111)-oriented crystal planes during near-equilibrium growth of Si using liquid phase epitaxy. An oxidized, (100)-oriented Si wafer with oxide-free seed lines oriented close to two orthogonal (110)-directions on the wafer surface serves as a substrate for epitaxial Si growth by Liquid Phase Epitaxy (LPE). This approach aims to fabricate a solar cell from the monocrystalline Si net and detach it from the wafer using a suitable etch [60]. Mitsubishi Electric Corp. developed the so-called VEST structure (Via-hole Etching for the Separation of the Thin film), where a Si layer is CVD-deposited and recrystallized on a SiO_2-covered mono-Si substrate. After the realization of this SOI structure, via-holes are etched with 100 μm diameters and a 1.5 mm distance between each other. The Si layer is subsequently detached from the wafer by HF-etching the SiO_2-intermediate layer through the via-holes. Solar cells were realized through this approach

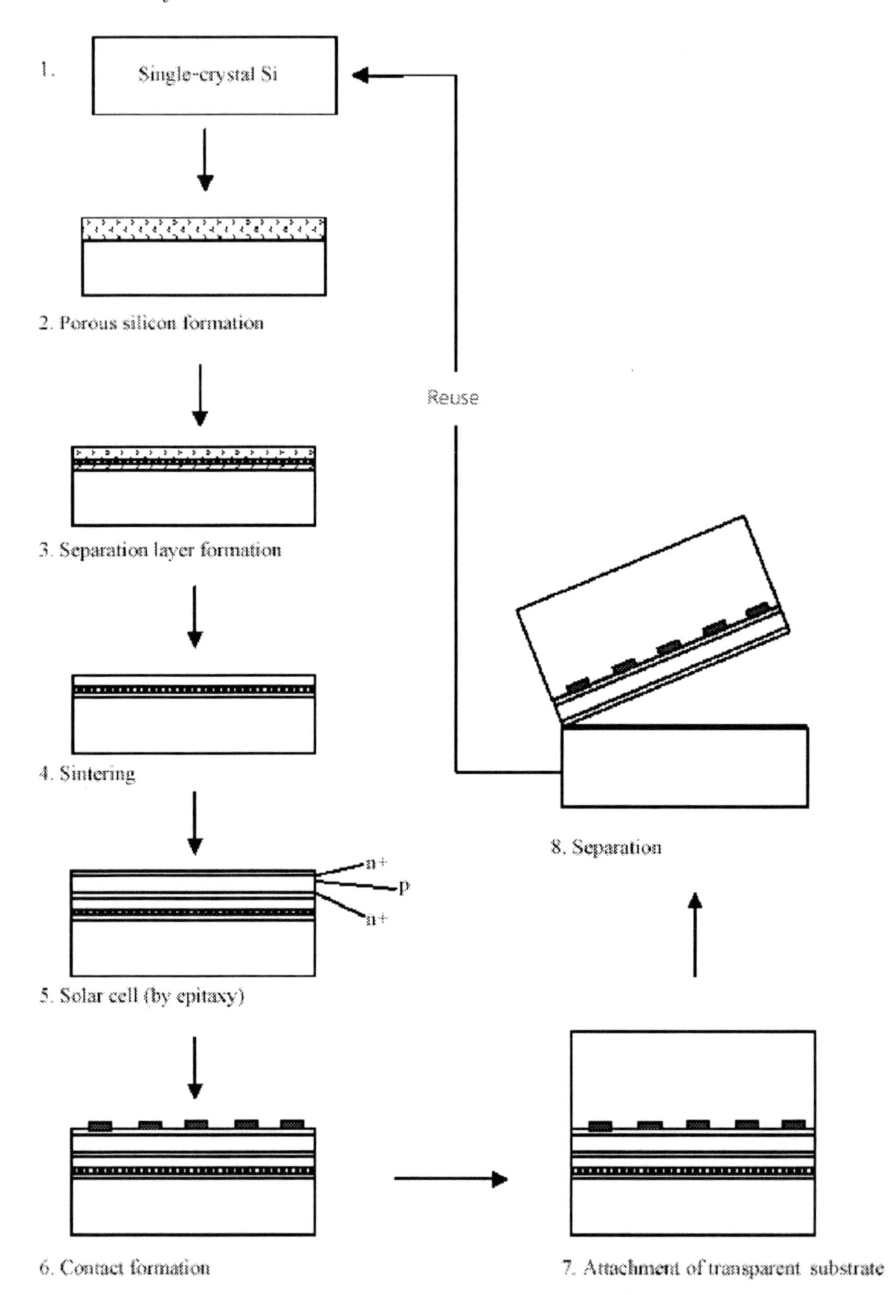

Fig. 4.7. Basic process sequence of lift-off technique based on layer of porous silicon

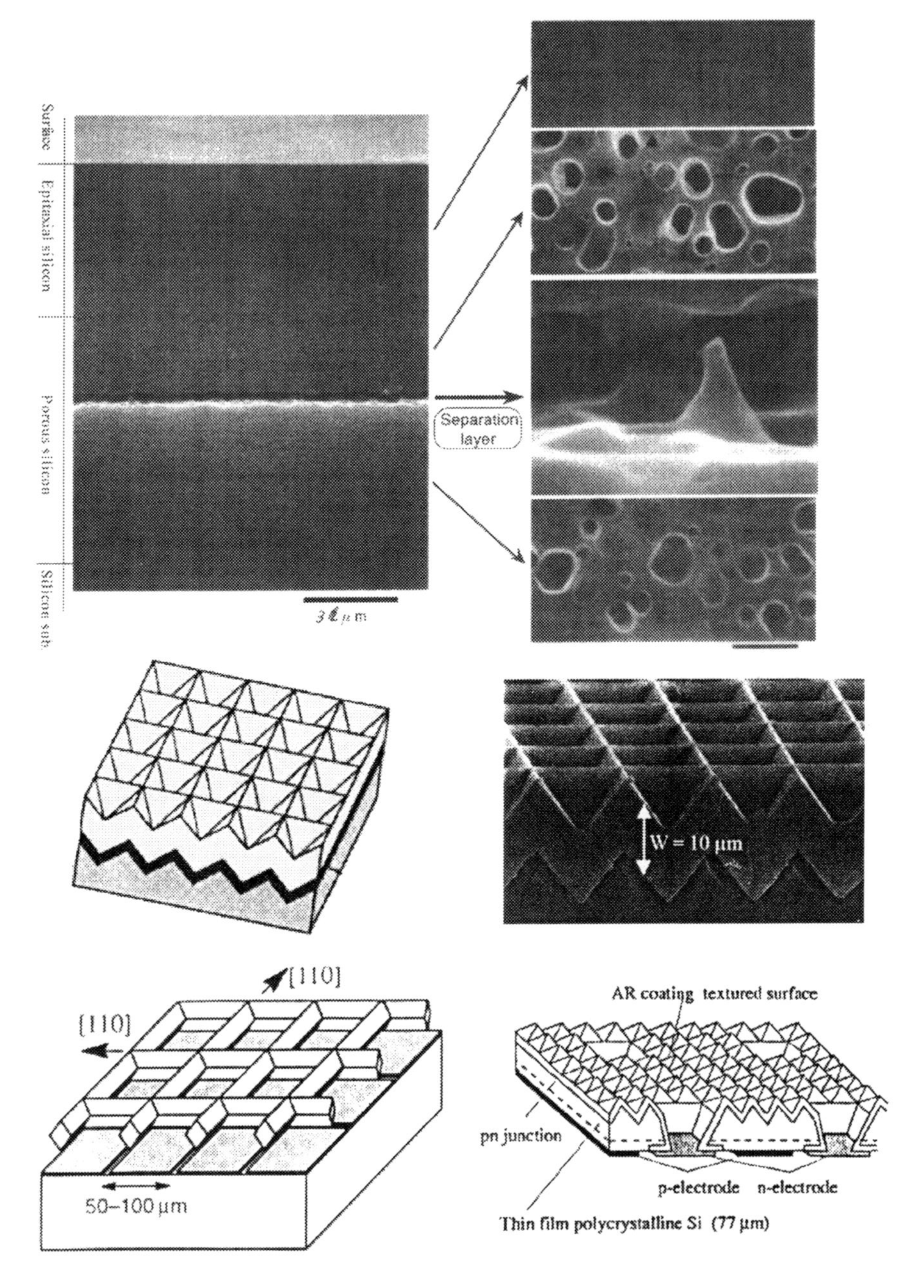

Fig. 4.8. Schematic of the most advanced transfer techniques of monocrystalline Si films: Separation by porous layer formation (**top**) [62], Ψ process (**middle**) [60], VEST process (**bottom right**) [61]

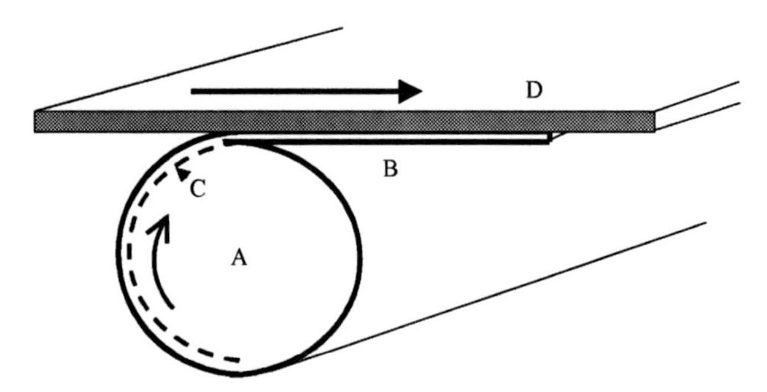

Fig. 4.9. Principle of large-area layer transfer by transferring the mantle of a cylindrical crystal [65]

with a thickness of 77 µm, an area of nearly 100 cm^2, and efficiencies of up to 16.0%. This is a remarkable success, especially due to the fact that the back contacts were screen printed on this fragile structure [61]. Sony Corp. pursues the epitaxial growth of monocrystalline Si film on the thermally annealed crystalline surface on top of a highly porous, buried Si level, as sketched in Fig. 4.8. The Si layer for the solar cell can be detached from the wafer due to the mechanically fragile separation layer underneath. Using a 12-µm-thick epitaxial film that was transferred on a plastic foil, a conversion efficiency of 12.5% on 4 cm^2 was achieved [62].

The Institute of Physical Electronics also relies on the epitaxial growth of the active Si layer on a so-called quasi-monocrystalline Si film. Again, a buried porous layer enables the separation of the solar cell from the wafer and the transfer of the processed cell to a foreign superstrate. Due to an excellent light-trapping scheme, efficiencies of up to 14.0% were obtained with this approach. Furthermore, it could be shown that the starting wafer can be used several times as a seeding wafer for the porous layer formation and the subsequent epi-growth [63].

The layer transfer process is just at the beginning of its development. It solves two problems of the present crystalline silicon technology:

- Material quality. Other than the deposition processes, it uses high-quality monocrystalline material. The use of thin layers can even lead to higher efficiency if proper design is employed.
- Efficient use of expensive single-crystal material. Only a thin surface layer is used for every transfer process, and the same wafer can be reused many times, as has already been shown.

A limitation compared to the large-area thin-film techniques to be described in the next chapter is that solar cell area is limited to the size of the wafer. Several concepts have been developed to improve this situation. One is to slice the crystals lengthwise. In this way, long rectangular substrates are obtained that, however, are of different widths [64].

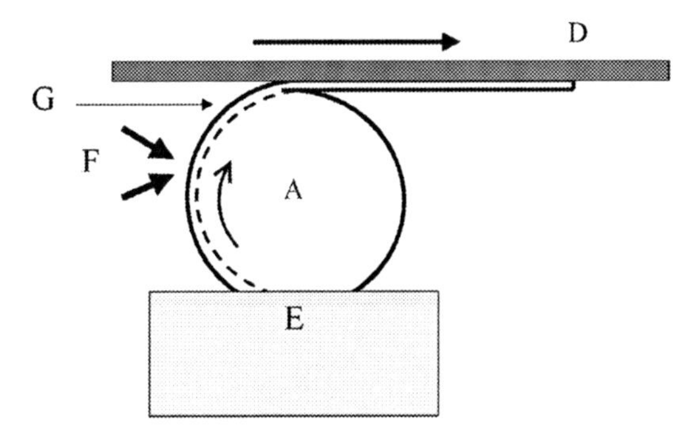

Fig. 4.10. Principle of continuous layer transfer

Another recently proposed approach can yield large areas in possibly even a continuous manner [65]. As shown in Fig. 4.9, the mantle surface of a cylindrical crystal can be peeled off. Before this step, the entire crystal is subjected to the electrolytic porosity treatment described above and possibly also to epitaxy. In Fig. 4.9, A is the cross section of the crystal, C is the separation layer, B is the detached film on the new substrate D. This concept can also be developed further into a continuous process, as shown in Fig. 4.10. In this figure, the crystal A rotates through an electrolytic bath E and is annealed by a linear heater F. At G an adhesive is supplied.

The potential gain in solar cell area is visualized by the following numbers: It has already been shown that for one transfer a thickness of 10 μm is consumed, including the sacrificial separation layer. If a crystal of 1 m length and an initial diameter of 30 cm is reduced to 20 cm, a total of 7800 m^2 active solar cell area is obtained. Assuming an efficiency of 15%, this crystal is converted into a 100 kW of solar cells, compared to 2 kW with today's wafer technology.

5 Other Materials, New Concepts, and Future Developments

5.1 Theoretical Efficiencies and Requirements for Solar Cell Materials

The maximum obtainable conversion efficiency of solar radiation to electricity has been studied very thoroughly. This efficiency can be derived in two ways: Thermodynamics and detailed balance [67]. The thermodynamic limit of a heat engine is given by the Carnot relation: $\eta = 1 - (T_2/T_1)$, where T_1 is the temperature of the heat source and T_2 that of the heat sink. The solar spectrum can be approximated by a black body of 5900 K. If the boundary conditions for terrestrial conversion are taken into account, a maximum efficiency of 85% results. The detailed balance principle is based on balancing the different particle fluxes in the solar cell and yields very similar results as the thermodynamic limit. It was introduced in an important paper by Shockley and Queisser [68]. Practical efficiencies today are far from the theoretical limit. The reasons can be easily understood:

- The solar spectrum is very broad and ranges from the ultraviolet to the near infrared, whereas a semiconductor can only convert photons with energy of the band gap with optimal efficiency. Photons with lower energy are not absorbed and those with higher energy are reduced to gap energy by thermalization of the photogenerated carriers. This situation can be improved by using several semiconductors in tandem cells, as we shall see below (Sect. 5.3.1).
- Sunlight arrives at the surface of the earth in a very dilute form compared to the way it leaves the surface of the sun. Direct sunlight can be concentrated by optical means, resulting in a much higher conversion efficiency.

We first look at a single semiconductor without concentration and consider its maximum efficiency. The curve of efficiency versus band gap goes through a maximum, as explained above and seen in Fig. 5.1. It can be noticed that silicon is not at the maximum but relatively close to it.

A large part of the book is concerned with silicon, although from solid state physics we know that silicon is not the ideal material for photovoltaic conversion. A very serious point is that silicon is an indirect semiconductor; valence band maximum and conduction band minimum are not opposite to

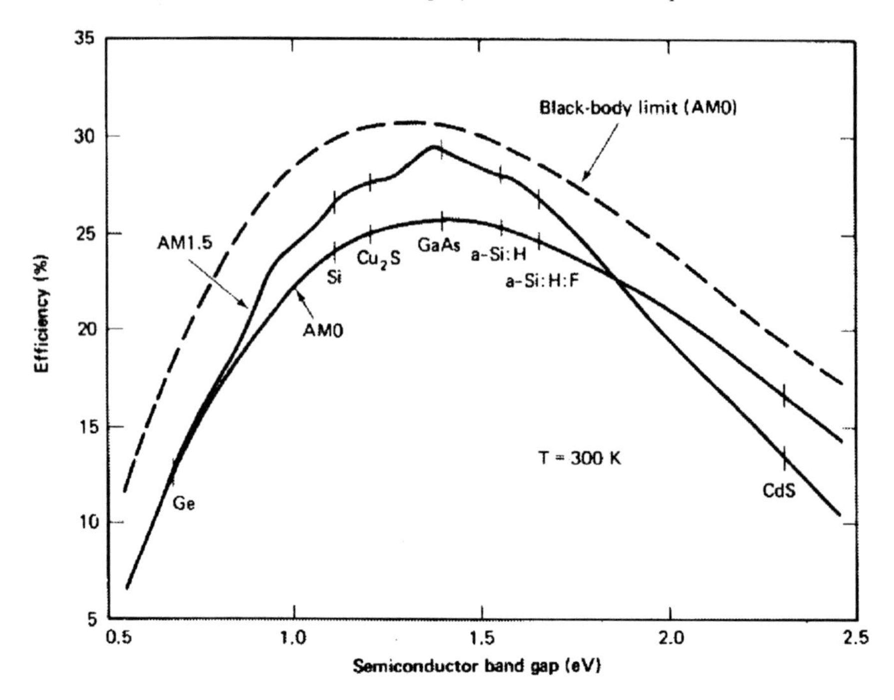

Fig. 5.1. Dependency of the conversion efficiency on the semiconductor band gap

each other in crystal momentum space, as described in Sect. 2.1. Light absorption is much weaker in an indirect semiconductor than in a direct semiconductor. This has serious consequences from a materials point of view: For a 90% light absorption, it takes only 1 µm of GaAs (a direct semiconductor) versus 100 µm of Si. The photogenerated carriers have to reach the pn junction, which is near the front surface. The diffusion length of minority carriers has to be 200 µm, or at least twice the silicon thickness. Thus, the material has to be of very high purity and of high crystalline perfection. In view of these physical limitations, it is quite surprising that silicon has played such a dominant role in the market. The main reason is that silicon technology had already been highly developed before the advent of photovoltaics and high quality material is being produced in large quantities for the microelectronics market at relatively low cost.

It is not surprising that a lot of effort has been going into the search for new materials. Requirements for the ideal solar cell material are collected in Table 5.1.

For the future of solar energy materials three scenarios can be envisioned:

- continued dominance of the present single crystal or cast polycrystal technology,
- new crystalline film Si materials of medium thickness either as ribbons or on foreign substrates,

Table 5.1. Requirements for the ideal solar cell material

	Si	CIGS/CdTe	a-Si
1. Band gap between 1.1 and 1.7 eV	+	+	+
2. Direct band structure	−	+	+
3. Consisting of readily available, non toxic material	+	−	+
4. Easy, reproducible deposition technique, suitable for large area production with good yield	−	+	+
5. Good photovoltaic conversion efficiency	+	+	−
6. Long-term stability	+	+	(+)

– breakthrough to mass production of true thin-film materials like a-Si or CIS or CdTe.

At this point, those scenarios have about equal probability. Even more likely is that two or three of them will coexist for a considerable period and that each technology will find its own market. From an overall point of view it can be considered an advantage that so many avenues exist that potentially lead to a low-cost solar cell. In this way, the likelihood of achieving this goal is greatly increased. In the long term, new concepts involving up and down conversion of quanta or new classes of materials like organic solar or thin-film tandem cells also are a possibility. They will be addressed at the end of this chapter.

5.2 Thin-Film Materials

As was pointed out in Chap. 2, thin-film materials are a way to reduce cost. In addition to the low materials consumption of thin-film materials, another advantage is that they can easily be connected in series in an integral manner on large-area substrates (see Fig. 5.4). Thus, entire modules are manufactured in the deposition process. This is advantageous economically, but it is also very demanding for the process technology, because large areas have to be processed without defects. Although their market share is low today, prospects for growth are considered good. One problem is that a number of different materials are being pursued, and it is not at all clear which one is the best choice. The most important materials and technologies will now be described.

5.2.1 Amorphous Silicon

History

The first publications on amorphous silicon (a-Si) relevant to solar cell fabrication appeared after the late 1960s [69–71]. The first paper at a photovoltaic conference was presented in the 12th IEEE PVSC (1976). Only five years later, the first consumer products appeared on the market. However, it took quite some time before the basic properties of the material were understood. More than twenty years have already passed since the first solar cell from amorphous silicon was reported by D. Carlson in 1976. The high expectancy in this material was curbed by the relatively low efficiency obtained thus far and by the initial light-induced degradation for this kind of solar cell (so-called Staebler–Wronski effect) [71, 72]. The efficiency drops when the cell is exposed to light. The degradation acts primarily on the fill factor and the short circuit current, whereas open-circuit voltage remains almost constant. Degradation can be reversed, but only by exposing the cells to a temperature of approximately 160°C. Even today the effect has not been fully explained. The most probable explanation is that the recombination of light-generated charge carriers causes weak silicon-hydrogen bonds to be broken in the amorphous material, thus creating additional defects that lower the collective efficiency and increase serial resistance. Much research work has been done to explain the cause of this effect. Today, it can at least be reduced by the use of technical measures. It has been established that after an initial degradation of 10 to 20%, the efficiency remains stable. Also, seasonal variations of efficiency have been observed: Efficiency drops in winter and recovers during summer due to annealing. The band gap of a-Si is better matched to the solar spectrum than that of crystalline silicon. Therefore, on a per watt basis amorphous cells have higher output integrated over one year. Today, a-Si has its fixed place in consumer applications, mainly for indoor use. After understanding and partly solving the problems of light-induced degradation, amorphous silicon begins to enter the power market. Stabilized cell efficiencies in the laboratory reach 13%. Module efficiencies are in the 6–8% range. The visual appearance of thin-film modules makes them attractive for facade applications.

Properties and Deposition Techniques

Amorphous silicon is an alloy of silicon with hydrogen. The distribution of bond length and bond angles disturbs the long range of the crystalline silicon lattice order and consequently changes the optical and electronic properties. The optical gap increases from 1.12 eV to about 1.7 eV.

The effective energy gap of amorphous Si is smaller than the effective band gap for light absorption. The edges of the valence and conduction band are not well defined, but exhibit a change in density of states. Charge carrier transport can occur at the mobility edge at energy levels that still have low absorption. This causes a material-inherent reduction of maximum obtainable efficiency. Furthermore, dangling bonds form deep levels in the forbidden gap of the material.

The basis for all deposition processes is silane (SiH_4) as precursor gas in a chemical vapor deposition (CVD) process. Typical deposition temperatures are below 500°C, otherwise no hydrogen is incorporated into the film. At low substrate temperature, pre-dissociation of the SiH_4 molecule is essential. The most commonly used method is plasma-enhanced chemical vapor deposition (PECVD). Besides a variety of designs of the plasma reactor (diode, triode configurations), a range of frequencies from radio frequency (RF) to ultra-high frequencies (UHF) is applied. Modifications like remote plasma reactors, where dissociation is spatially separated from deposition or dissociation by an electron cyclotron resonance reactor (ECR) or a hot wire (HWCVD), have an influence on film quality.

The challenge is to produce a material with well-defined disorder, which is a contradiction in itself. Therefore, the light-induced defect modification cannot be completely avoided. It turns out that the stability of the material is related to the hydrogen content and not so much the process itself. The best quality is reached by diluting SiH_4 with hydrogen. It appears that a-Si:H has reached a state in which further improvement of the material itself is not very likely.

Improvements of solar cells now have to rely mostly on device design. Manufacturability may be improved by increasing the deposition rate while maintaining sufficient material quality. Hydrogen dilution allows control of the structure of the film. At very high dilution ($< 90\%$), a transition to the microcrystalline (μc) state of the material occurs. The band gap of the μc material approaches that of crystalline silicon. Due to internal scattering, the apparent optical absorption is nearly one order of magnitude higher compared to the crystalline material. At the low doping levels in the intrinsic layer, and due to passivation by the high hydrogen content, grain boundaries do not have a strong influence on device performance.

Solar Cell and Module Properties

The mobility of charge carriers in amorphous silicon is generally quite low so that the collection of photogenerated carriers has to be supported by an internal electrical field. Furthermore, defect formation is related to the recombination process. The typical structure and corresponding band diagram of an amorphous silicon solar cell is depicted in Fig. 5.2.

In order to create a high field in the intrinsic layer of the pin structure, the cells have to be thin, of the order of a few hundred nanometers. By applying proper light trapping schemes, thin single junction cells can be produced with high stabilized efficiency [73].

Therefore, the general strategy to improve the stabilized efficiency of the devices is to use a stack of two or even three cells. It allows the reduction of thickness of the single cells, increasing the electric field and hence improving carrier collection. The relatively low substrate temperature for the deposition of the films has the advantage that the underlying layers are not affected by the subsequent deposition steps. Therefore, production of these stacks in a monolithic structure does not add too much deposition time, so additional processing costs are not substantially higher than for single cells. Introducing

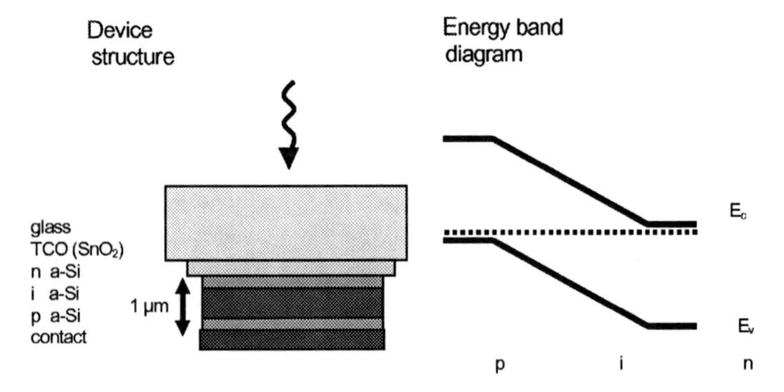

Fig. 5.2. Schematic structure of an amorphous pin solar cell and corresponding band structure

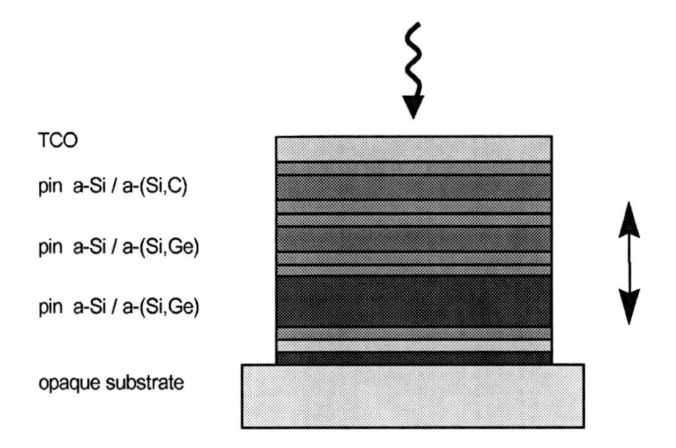

Fig. 5.3. Structure of a substrate-type amorphous silicon tandem solar cell

cells with different bandgaps in the stack results in a “real” tandem cell (see Sect. 5.3.1) that can make better use of the solar spectrum and at the same time improve the stabilized efficiency of the devices. The band gap can be increased by alloying with carbon. Solar cells are produced in both substrate and superstrate configurations. Both have their own advantages. Superstrate cells are deposited on transparent conductive oxide (TCO) coated glass substrates. Substrate cells allow the use of flexible metal or polymer foils. However, monolithic series connection requires an insulating substrate. Therefore, cells on stainless steel substrates are usually cut into single cells and interconnected mechanically. The schematic view of a triple tandem cell on a stainless steel substrate produced by United Solar is shown in Fig. 5.3.

One decisive advantage of the a-Si cell is that the necessary serial connection of cells can take place simultaneously during manufacture. As shown in Fig. 5.4, an entire layer of TCO_3 is first deposited onto a glass substrate, and then a stripe pattern is created, for example, by a laser beam. Then the

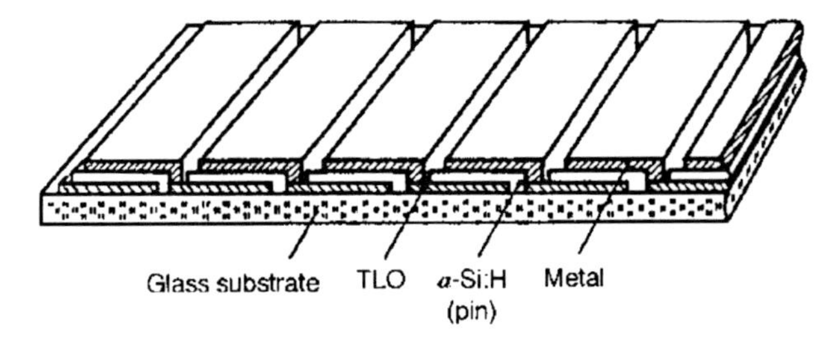

Fig. 5.4. Integrated serial connection of a-Si solar cell

a-Si solar cell structure is deposited in a reactor, the cells are structured, again using a laser, so that the subsequent evaporated metal has contact to the TCO on the back of the glass substrate. Finally, the evaporated metal coating must be separated at a suitable offset. It is thus clear that in this case five cells are connected in series. In this manner, the output voltage can be adapted to the load. This effect is decisive in making a-Si solar cells dominate almost exclusively the small output market (clocks, pocket calculators, etc.).

Status and Future Prospects

Two recent top values for the efficiency of a-Si–based solar cells are presented in Table 5.2. The devices consist of either two or three cells in a stack. The second line gives data for the microcrystalline device fabricated by IMT Neuchatel based on amorphous semiconductors [74].

Different approaches to a-Si module manufacturing are listed in Table 5.3. Module manufacturing of amorphous silicon follows mainly two lines, namely, flexible foil substrates or glass substrates. The flexible substrates offer new

Table 5.2. Efficiencies (initial and stabilized) reached with laboratory scale devices based on amorphous semiconductors

Cell/type	eff. (init./stab. [%])	Manufacturer
a-Si/a-SiGe/a-SiGe	15.6/13.0	United Solar
a-Si/μc-Si	13.1/10.7	IMT Neuchatel

Table 5.3. Different approaches to a-Si module manufacturing

Module type	Size/prod.volume	Company
a-Si, a-Si, a-SiGe, flexible stainless steel	stainless steel foils, 5 MW	United Solar
Double a-Si, a-SiGe, glass	$0.46\,m^2$, 10 MW	Solarex
a-Si, a-Si, flexible, polymer	$40 \times 80\,cm^2$	FUJI
a-Si, glass	$91 \times 45\,cm^2$, 20 MW	Kaneka
a-Si, a-Si, glass	$0.6\,m^2$, 1 MW	ASE, PST

applications. However, monolithic interconnects of these cells are a challenging task.

Typical substrate sizes are on the order of 0.5 to 1 m^2. Stabilized efficiencies of large-area modules are about 6 to 8% for multijunction cells and optimized single junction cells. The near-term goal is to reach 8% stabilized efficiency on large areas on a production level. United Solar modules contain subcells that are laminated to a single cover sheet. The very large module of EPV consists of 4 submodules laminated to a single cover glass. The setup of the production of modules on the multi-Megawatt level is presently going on at United Solar and Kaneka. Other companies operate pilot production lines for large-area modules and are planning to upscale the manufacturing lines.

Progress in amorphous silicon development is related to the following areas.

- General understanding of the material and stability issues. The understanding of the mechanism of the Staebler–Wronski effect being basically an intrinsic property of the material that adjusts its defect equilibrium according to the operating conditions.
- Improvement of the process. Hydrogen dilution of silane in the plasma discharge and very high-frequency plasma deposition lead to improved material properties also at higher deposition rates.
- Understanding the structural properties of the material at the borderline between the amorphous and crystalline states, so-called "protocrystalline" Si.
- Optimize cell designs by introducing buffer layers, alloy and doping gradients, etc. to increase cell efficiency and reduce the effect of initial degradation.
- Optimized stacked cells and tandem devices lead to the highest stabilized efficiencies.

The "classical" approach uses an alloy with Ge to reduce the band gap in the bottom cell(s) to about 1.5 eV. However, the corresponding process gas GeH_4 substantially contributes to the cost of the module, so an alternative development for the low band gap cell is advisable. One promising solution is the use of microcrystalline silicon. IMT Neuchatel demonstrated such a solar cell based on microcrystalline silicon. The stable efficiency of over 8% reached till now may be low for a stand-alone thin-film cell. However, in the so-called "micromorph concept", i.e., a micromorph tandem concept, this cell could replace the Si, Ge bottom cell. So far the open-circuit voltage of such microcrystalline cells is below 0.5 V. The electronic quality of the material depends strongly on the purity of the gases used in the process. Further investigations are needed to identify the limiting factors for solar cells based on this material.

The deposition of amorphous silicon at very low substrate temperatures, below 100°C, still yields devices with reasonable efficiencies.

Amorphous silicon is firmly entrenched in the consumer product market, but because of the strong expansion of the grid-connected market in recent years, crystalline silicon has gained market share. The present situation still is characterized by a large uncertainty about the future of a-Si. BP-Solar, who took over Solarex, have recently closed the amorphous plant that was set up and developed over many years by Solarex. BP intends to concentrate on the crystalline silicon technology.

5.2.2 Copper Indium Diselenide and Related Compounds

History

A very challenging technology is based on the ternary compound semiconductors $CuInSe_2$, $CuGaSe_2$, $CuInS_2$, and their multinary alloy $Cu(In,Ga)(S,Se)_2$ (in further text: CIGS). The first results of single crystal work on CuInSe (CIS) [75] were extremely promising, but the complexity of the material looked complicated as a thin-film technology. Pioneering work by Kazmerski [76], however, showed immediate success. It became evident that CIS process technology is very flexible with respect to process conditions. Establishing a well controlled system for multisource coevaporation by Boeing soon made the CIS cell the frontrunner with respect to thin-film solar cell efficiencies. ARCO Solar developed in the mid-1980s a fabrication technology that is better adapted to current thin-film processing, namely, sputtering of metal films with a subsequent selenisation. In later developments, the addition of Ga and sulfur helped increase the efficiency [77].

Material Properties and Deposition Techniques

The phase composition of the ternary compound is mostly described by the pseudobinary phase diagram of the binary Cu_2Se and In_2Se_3 phase. Recent investigations showed that single phase chalcopyrite $CuInSe_2$ only exists at a small copper deficiency. In Indium rich films, defect chalcopyrite phases segregate. Nevertheless, the electronic properties of the compound are not very much affected by deviation from stoichiometry. A plausible explanation was derived from theoretical calculations. Cu vacancies form together with In interstitials a neutral defect complex:

$$2(V_{Cu})^- + (In_{Cu})^{++}.$$

The energy levels of this complex lie in the valence or conduction band. Furthermore, these chalcopyrite semiconductor compounds form a Cu-depleted defect layer at the surface. This behavior has important consequences for junction formation because it forms a type inversion at the surface of the p-type film.

The importance of sodium for the quality of CIGS films became more and more evident. Sodium not only improves crystallization of the film, but

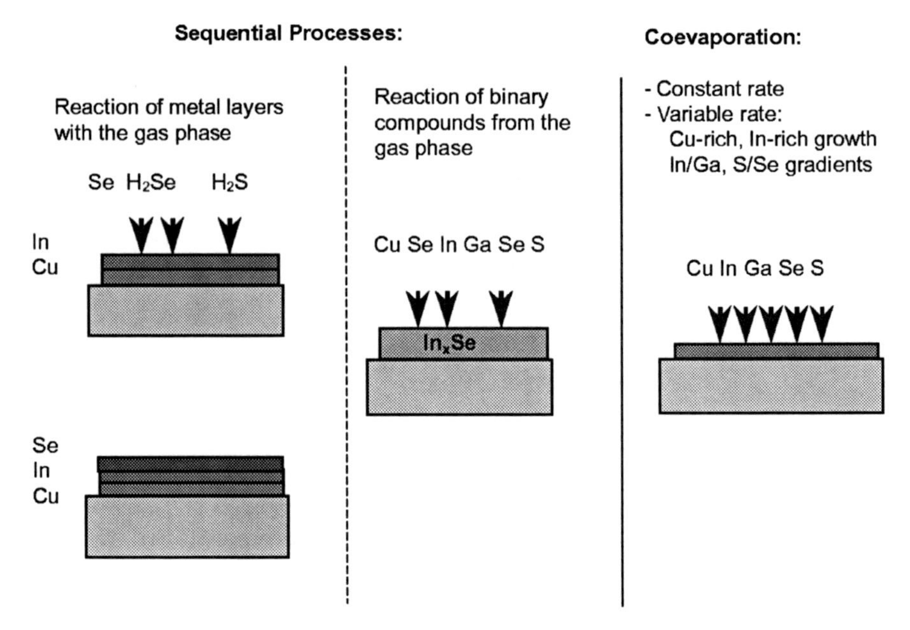

Fig. 5.5. Comparison of approaches: selenization of metallic precursors versus coevaporation

also increases conductivity. The mechanism is still not known. There is much evidence that sodium is incorporated at grain boundaries or defects. Concentration of Na in the bulk must be very low.

For the large-scale fabrication of these rather complex compounds, two different approaches or even philosophies exist for the deposition of the absorber layer.

- Deposition of precursor layers and a subsequent treatment or annealing in H_2Se vapor. The deposition is divided into several simple steps. These processes can use "off the shelf" equipment as much as possible and therefore need in principle only the process development. However, this strategy has had limited success till now. Furthermore, it limits the composition of the absorber layer to low bandgap In-rich material.
- The process that yields the best performance on small-area laboratory-scale devices is coevaporation. This procedure gives full flexibility in device optimization. However, it is a real challenge for the engineers to design appropriate evaporation sources while somehow achieving the guarantee of high-efficiency upscaling.

Figure 5.5 and Table 5.4 illustrate the different methods for CIGS film deposition. Careful analysis and future experience have to show which option is most suitable for upscaling. The addition of several simple steps could be less economic than one sophisticated but efficient process. Deposition speed

Table 5.4. Deposition of CIGS films for module manufacturing

Deposition method	Company
Metal precursor + H_2Se selenization	Siemens Solar Industries (SSI), Showa
Metal/Se precursor, anneal	Siemens Solar
Metal precursor no vacuum	ISET, Unisun
Coevaporation	ZSW, Global Solar

of in-line coevaporation can be as high as 5 cm/min with the presently known parameters and source design. This means a module with the length of 1 m could be deposited in 20 min.

The cells produced by the selenization process contain gallium only toward the back of the cells, because of the principal difficulty of incorporating this element homogeneously in the film during this process. Therefore, the band gap toward the surface is widened by introducing sulfur, forming a graded band gap structure.

The selenization technique is now the basis of the first pilot production and market introduction of CIS modules by Siemens Solar. Aperture area efficiencies of over 12% make these modules attractive for power applications. Evaporation processes for compounds consisting of four elements like $Cu(In,Ga)Se_2$ on large areas have been considered unsolvable problems. However, based on the long-term experience at IPE dating back to the CdS deposition, the Center for Solar Energy and Hydrogen Research (ZSW) demonstrated coevaporation processes on 60-cm-wide glass substrates. Global Solar is depositing the cell in a continuous process on a stainless steel or polymer web. Since vacuum deposition systems require high investments, other methods like particle deposition are under investigation.

Solar Cell and Module Properties

CIGS module fabrication has the same advantageous features of thin-film fabrication processes as the other thin-film solar cell materials. The typical device structure shown in Fig. 5.6 is based on a soda lime glass substrate, which often also serves as a source for the doping of the CIGS films with sodium. Typical solar cells are deposited on Mo-coated glass substrates at a substrate temperature above 500°C. The heterojunction is formed by chemical deposition of a thin CdS layer from a solution containing Cd ions and thiourea as a sulfur source. This process has many intrinsic advantages, so it is difficult to replace it by either another Cd-free compound or a more convenient gas phase process. The most efficient devices and first large-scale productions of modules still rely on the CdS layer. However, research in many labs concentrates on the replacement of the CdS layer. One effort in Japan includes a wet-deposited Zn(S,OH) compound already in pilot production.

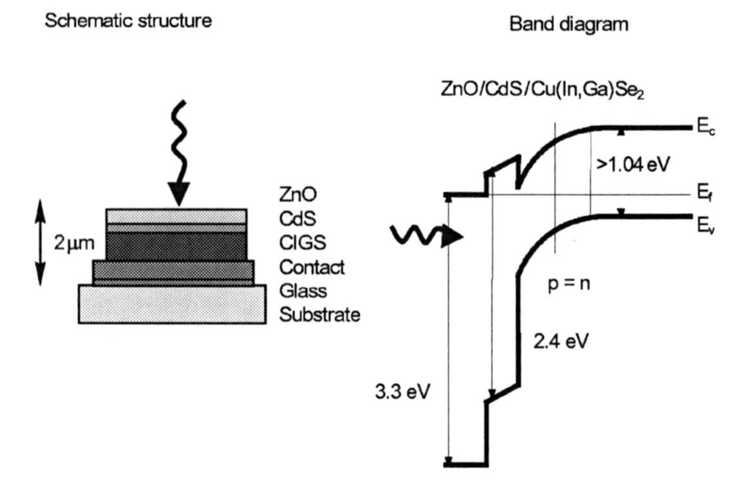

Fig. 5.6. Schematic structure of a CIGS-based solar cell

Attempts to realize superstrate cells, as with CdTe or a-Si, are hampered by the high deposition temperature for the CIGS film, resulting in a non-controllable interdiffusion at the heterojunction. Therefore, the efficiency of these devices is limited to 10%.

Status and Future Prospects

Very high efficiencies approaching 19% have been reported for laboratory scale devices [78]. This result has been obtained for this material system by empirical optimization of process parameters. Due to the high flexibility of designing compounds with defined properties in this material system (e.g., band gap grading), even more improvements can be expected in the near future. Table 5.5 summarizes the best values of solar cell performance obtained for this material system.

A further challenge is to realize high voltage devices on the basis of the ternary chalcopyrite compounds $CuInS_2$ and $CuGaSe_2$. The efficiency achieved with these wide energy gap materials are still considerably lower than those of the low gap $Cu(In,Ga)Se_2$. This limitation is mostly due to the low open circuit voltage compared with the band gap. However, improved understanding of the materials, especially the surface and junction properties, may help to further develop the devices in the near future and open new perspectives of ternary thin-film photovoltaic devices. Further developments are directed toward replacement of the CdS buffer layer and probably the reduction of film thickness or even replacement of In and Ga as rare elements. Pilot production of CIS modules is currently under way at Shell (formerly Siemens Solar) in the U.S.A. and Würth Solar in Germany.

Table 5.5. Performance of CIGS cells and modules (independently confirmed at * NREL, + ISE/Fraunhofer)

Process	Lab. cell efficiency [%]	Module eff. [%] / area [cm^2]	Laboratory/company status
Selenization of precursor	> 16	12.1 / 1 × 4 ft*	Siemens, pilot production
Metal films		14.7 / 18^+	
Coevaporation/ sequential	18.8		NREL
Evaporation			
	17.2	13.9 / 90^+	IPE
	16.2	12.7 / 800^+	ZSW, laboratory production; Würth Solar, pilot production
		9.6 / 135*	EPV
	11.5	5.6 / 240	Global Solar
		16.8 / 19^+	Angstroem Solar Center
Non-vacuum processes	> 11	8 / 74	ISET, Unisun

5.2.3 Cadmium Telluride

History

Thin-film solar cells based on CdTe are the cells with the longest tradition, but they are really not "of age." After a long development, they arrived at cell efficiencies of 16% and large-area module efficiencies of over 10%. There is no reason why they should not improve further in the foreseeable future. The history of CdTe-based cells is a story about an adventurous search for the appropriate structure on a track full of obstacles and traps. As in the case of CdS, first attempts have been made with CdTe single crystals. At the RCA Labs, indium was alloyed into n-type CdTe crystals, resulting in an alloy-type pn-junction with 2.1% conversion efficiency. At the same time, CdTe-cell efficiencies of even 4% were published in the USSR and submitted for publication even earlier than the RCA paper. Cusano reported on the first thin-film cell; he used a structure similar to the CdS cells, namely, a p-Cu_2Te/n-CdTe heterojunction, and succeeded in obtaining 6% efficiency, an excellent value for the first trial at that time [79]. For nine years after the work reported by Cusano there was not much to report about this material. In this interval, CdTe was used as p-type material in conjunction with n-CdS, not only for testing pn-heterojunctions, but also for producing graded-gap junctions by interdiffusion of the materials. Achieving not more than 6% efficiency, the researchers saw the dominant problems of CdTe solar cell development:

- the difficulty of doping p-type CdTe;
- the difficulty in obtaining low-resistance contacts to p-type CdTe, and the recombination losses associated with the junction interface.

In the years to come, more than ten different types of solar cell structures with CdTe as the absorber will be tested. The borderline of 10% efficiency was crossed in 1982, astonishingly again with a n-CdS/p-CdTe heterojunction made in the Kodak labs [80]. Oxygen was present during CdTe- and CdS-layer deposition, acting as effective p-dopant in CdTe. Furthermore, the cell featured only a very thin CdS layer that improved the blue response. Very thin CdS layers acting only as a buffer layer in the heterostructure with the transparent conductive oxide – see also $CuInSe_2$ cells – were the key to further improvements. The research groups at the National Renewable Laboratory and at the University of South Florida (USF) [81] are those having pushed the efficiency to the range of 16% .

A large number of deposition methods have been applied to CdTe, resulting in high-quality layers and high-efficiency cells in economic production. Close-spaced sublimation (CSS) is the most popular technique for obtaining the highest efficiencies, spraying or screen printing are techniques with high economic potential. It is remarkable that the highest efficiency CdTe PV devices are fabricated from polycrystalline rather than single crystalline CdTe. A fundamental question therefore is, Why are polycrystalline thin-film PV devices more efficient than their single crystal counterparts? A possible answer is that grain boundaries enhance the collection of photogenerated minority carriers! There are indications that an electron barrier exists at grain boundaries that makes devices fabricated from II–VI compounds generally less sensitive to grain boundaries than devices made from III–V or group IV materials.

Material Properties and Deposition Techniques

CdTe is a nearly ideal material for thin-film photovoltaics because it combines several advantageous properties. Besides an optical band gap close to the optimum for solar energy conversion, it is very easy to handle thin-film deposition processes. Therefore, many efforts were and still are directed toward the large-scale fabrication of modules. Laboratory cells reach efficiencies above 16%.

The congruent evaporation of the compound, i.e., the evaporation of stoichiometric CdTe, results in a stoichiometric composition of the vapor. Above a substrate temperature of a few hundred degrees Celsius the composition is self-stabilizing. High quality material can be deposited at very high rates ($> 1 \mu m/min$) at substrate temperatures of 450–600°C. Because of the tolerance of the material to defects and grain boundaries, simple processes such as electrodeposition and screen printing are possible, and these processes are a good prerequisite for large-scale production. The highest quality material and hence the highest efficiencies are obtained with close-spaced sublimation

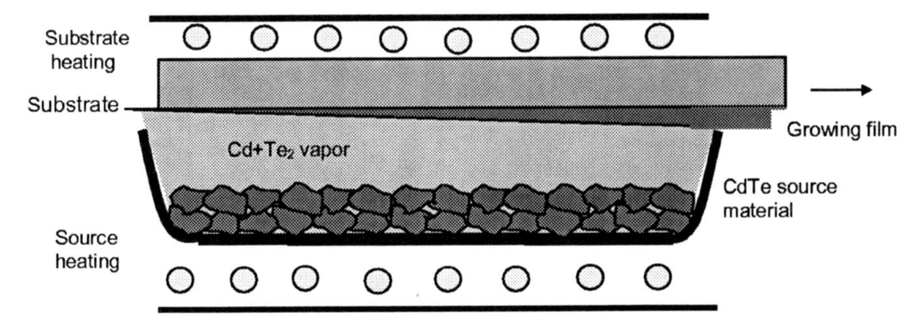

Fig. 5.7. Schematic view of a reactor for the continuous deposition of CdTe by close-spaced sublimation

Table 5.6. Approaches to CdTe module manufacturing. All companies fabricate modules with sizes $> 0.5\,m^2$

Status, companies	Deposition technique
Production started (First Solar, Matsushita, ANTEC)	Close-spaced sublimation
Pilot production abandoned by BP Solar	Electrodeposition
Matsushita	Screen printing

(CSS), a modified evaporation process, where substrates and sources are very close together with a relatively small difference in temperature so that the film growth occurs close to equilibrium condition (Fig. 5.7).

The different methods for CdTe deposition in Table 5.6 give an impression of the flexibility of the material with respect to manufacturing. In-line processes with high deposition speed (CSS) or processes where large batches can be handled (electrodeposition) have been developed. Some companies involved in CdTe development plan upscaling and a multimegawatt production in the near future.

Solar Cell and Module Properties

The active layers of a CdTe-based solar cell are deposited on TCO- (SnO_2 or indium tin oxide) coated glass like in a-Si superstrate cells. High efficiency cells use very thin chemically deposited CdS (see also CIGS-based cells). In-line processes by close-spaced sublimation usually include CdS deposition. In this case, the CdS layer has to be thicker for a good coverage of the substrate. Therefore, losses in the CdS window are increased. The schematic structure of the typical CdTe-based solar cell is shown in Fig. 5.8.

CdTe–CdS solar cell fabrication includes the following major technical issues:

- junction formation,
- crystallization of the films,
- formation of stable back contacts.

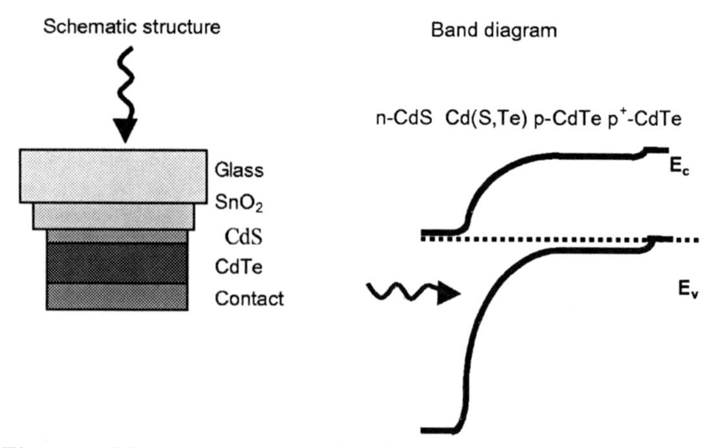

Fig. 5.8. Schematic structure of a CdTe-based superstrate solar cell

The fabrication of the heterojunction is one of the keys to efficient devices. Due to the intermixing of the CdS and the CdTe layers, not an abrupt heterojunction but a graded gap structure is formed. Interdiffusion of the CdS and CdTe layers during fabrication or after a $CdCl_2$ treatment determines the electrical properties of the device. The $CdCl_2$ treatment of the CdTe film also enhances the size of the crystallites in the polycrystalline films. The fabrication of stable ohmic contacts with the wide gap p-type semiconductor CdTe is a prerequisite for stable modules. Complex recipes for contact formation include chemical treatments for creating a Te-rich surface combined with the deposition of Sb, Carbon, or Cu.

Status and Future Prospects

At present there are several attempts to set up production lines for CdTe-based modules with capacities in the multimegawatt range. The activities of the different companies are listed in Table 5.6. The area of monolithic thin-film modules approaches one square meter. The monolithic thin-film module with the world's highest power output has been produced based on CdTe. At BP Solarex a monolithic module with a power output exceeding 70 W has been fabricated. In this case, the CdTe layer has been electrodeposited. BP Solarex, however, announced the close of their CdTe plant along with that of the amorphous line. Another blow to this technology is the insolvency of Antec in Germany due to the bankruptcy of its mother company Babcoc. A new investor is now continuing production. First Solar attempts to connect solar cell production to a float glass line. Reaching 20% device efficiency might be possible. The future has to show how much sophisticated processing can be translated to large-area high-throughput production.

A nontechnical problem associated with CdTe is the acceptance in the marketplace, because Cd and to a lesser extent Te are toxic materials, although the compound is quite stable and harmless.

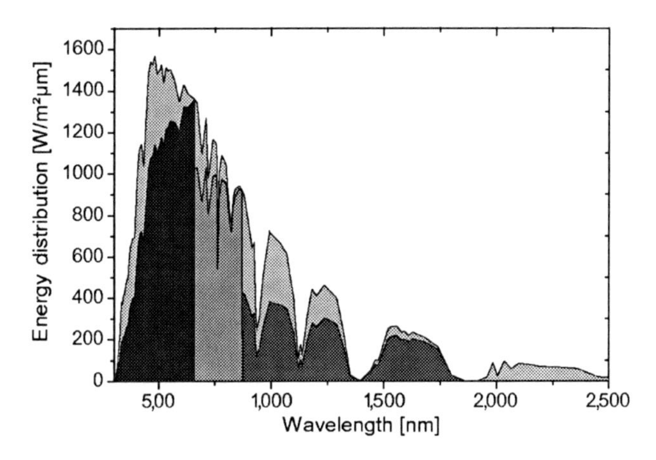

Fig. 5.9. Conversion of solar spectrum by three cells in series. Each cell converts only part of the spectrum with optimal efficiency

5.3 Other Materials and Concepts

5.3.1 Tandem Cells, Concentrating Systems

The efficiency of solar cells can be significantly increased by stacking several cells with different band gaps, such that the gap energy decreases from the top. That way each solar cell converts part of the solar spectrum at maximum efficiency. The arrangement has to be as follows: The top cell consists of a wide gap semiconductor. It converts the short wavelength part of the spectrum and transmits the other part to the cells below and so on. Figure 5.9 shows the solar spectrum and its subdivision by three solar cells in a tandem arrangement. (Although tandem originally referred to two cells, it is now also used for arrangements with more cells.) Two cells in series connection have a maximum theoretical efficiency of 41.9%, and with a larger number of cells 50% can be exceeded. Such tandem arrangements can be realized either with a sequence of thin films, as has already been demonstrated with amorphous silicon, or they can be incorporated into concentrating systems. Figure 5.10 shows a monolithic concentrator cell consisting of three sub-cells based on III–V compounds and on the appropriate splitting of the solar spectrum. A technological problem is that rectifying junctions exist between the cells. They have to be bridged by tunnel junctions. The figure illustrates the complexity of these cells. A problem connected with tandem cells is that with series connected cells an equal number of photons has to be absorbed (and converted) in each cell. This can only be accomplished, if at all, at a given spectral distribution like AM 1.5. Since this distribution changes under terrestrial conditions, spectral mismatch will occur. An obvious but complex solution is to provide separate contacts to each cell. Such an arrangement with mechanically stacked cells is shown in Fig. 5.11.

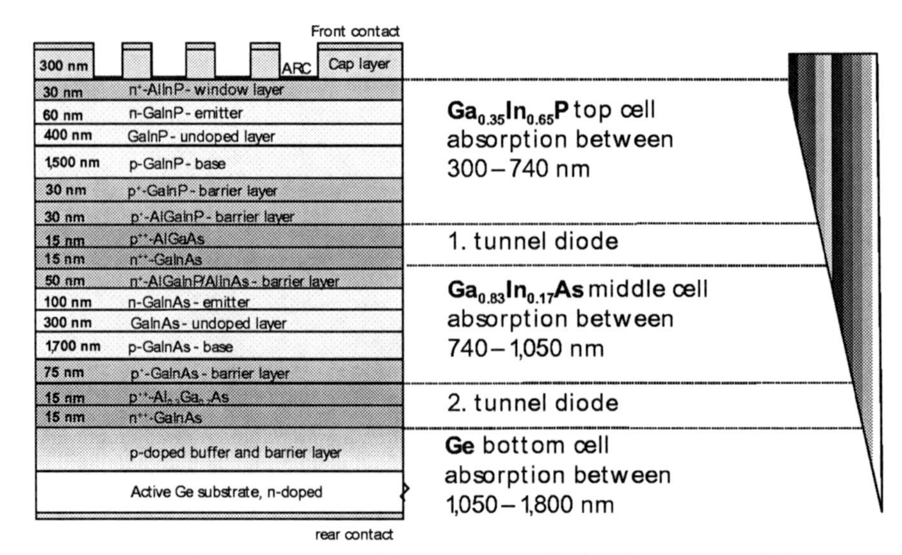

Fig. 5.10. Triple-junction monolithic tandem cell. At **right**, spectral absorption versus depth

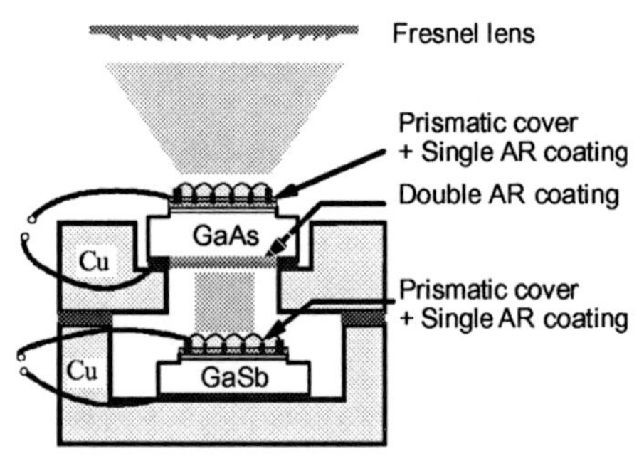

Fig. 5.11. Mechanically stacked tandem cell with light concentration

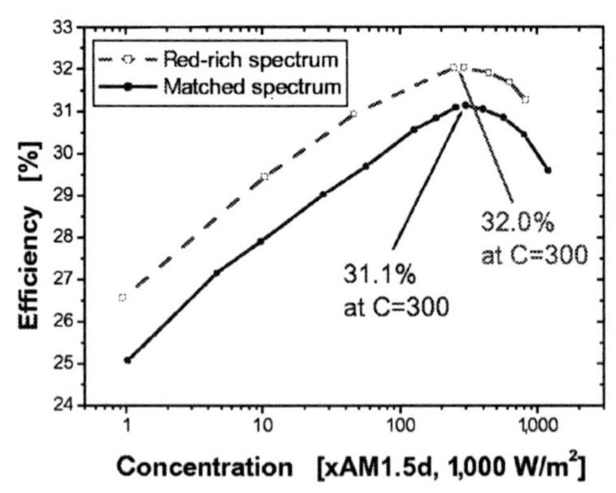

Fig. 5.12. Efficiency versus light concentration for two different spectral distributions

Concentration of sunlight is an approach suitable for III–V cells in order to reduce the cost of conversion. On the one hand, with III–V cells the highest efficiencies can be obtained; on the other, they are very expensive. Concentration is only feasible with direct sunlight, which can be concentrated by optical elements tracking the sun. If the concentration factor is very high, then the cost of the solar cell is only a small part of systems cost and, therefore, the solar cells can be expensive as long as efficiency is very high [82]. For this purpose, even high cost III–V compounds are possible, including tandem systems. Concentration of radiation also leads to increased efficiency, since V_{oc} is proportional to the logarithm of light-generated current density. This effect is demonstrated in Fig. 5.12. The efficiency rises and goes through a maximum, because at very high concentration series resistance comes into play. The record conversion efficiency for a solar cell is held by a tandem cell with 38% at high light concentration[1]. Tandem cells are now in the process of being introduced into the space (satellite) market.

5.3.2 Dye-Sensitized Cells

Nano-crystalline dye-sensitized solar cells are based on the mechanism of a fast regenerative photoelectrochemical process. The main difference of this type of solar cell compared with conventional cells is that the functional element responsible for light absorption (the dye) is separated from the charge carrier transport itself. In the case of the n-type semiconductor TiO_2 (band gap 3.2 eV) (Fig. 5.13), this results in a working cycle starting with the dye excitation by an absorbed photon at the TiO_2-electrolyte interface and an electron injection into the TiO_2. The injected electrons may migrate to the front electrode (a transparent TCO glass) and can be extracted as an external current. The dye is subsequently reduced by a redox electrolyte, based on an organic solvent and the redox couple iodide/triiodide. The redox electrolyte also accomplishes the charge transport between the counter electrode (also a TCO glass) and the dye molecules. For a low-resistant electron transfer, the counter electrode is covered with some Pt, which acts as a catalyst for the redox reaction.

It could be shown that only dye molecules directly attached to the semiconductor surface are able to efficiently inject charge carriers into the semiconductor with a quantum yield of more than 90%. As the overall light absorption of a dye monolayer is only small, this limits the photocurrent efficiency with respect to the incident light to a value well below 1%. This mechanism could be evaded by the preparation of titanium dioxide electrodes with a nanoporous morphology resulting in a roughness factor of about 1,000. After the announcement of surprisingly high efficiencies by M. Grätzel and coworkers in the early '90s [83], this type of solar cell is under reinforced

[1] Y. Yamaguchi, announced at the EU-PV Conference, Paris, 2004.

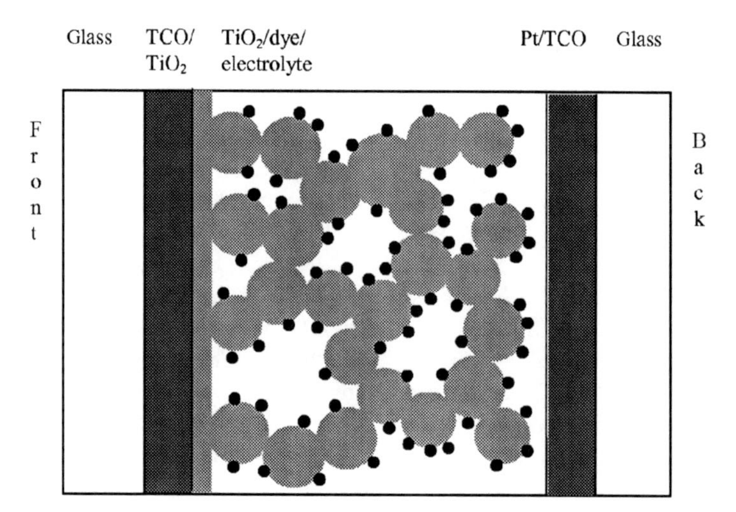

Fig. 5.13. Principle of the dye-sensitized solar cell

development aiming at large-area and low-cost solar. After the first experimental success, it took nearly a decade before the first quantitative models were established that link the material parameters of the constituents with the electrical performance of the whole cell such as the I–V characteristic and the spectral response.

The major advantage of the concept of dye sensitization is the fact that the conduction mechanism is based on a majority carrier transport as opposed to the minority carrier transport of conventional inorganic cells. This means that bulk or surface recombination of the charge carriers in the TiO_2 semiconductor cannot happen. Thus, impure starting materials and a simple cell processing without any clean room steps are permitted, resulting in promising conversion efficiencies of 7–11% and the hope of a low-cost device for photoelectrochemical solar energy conversion. However, impure materials can strongly reduce the lifetime of the cells.

The most important issue of the dye-sensitized cells is the stability over the time and temperature ranges that occur under outdoor conditions. Although it could be shown that intrinsic degradation can considerably be reduced, the behavior of the liquid electrolyte under extreme conditions is still a problem. For a successful commercialization of these cells, the encapsulation/sealing, coloration, and electrolyte filling have to be transferred into fully automated lines, including the final closure of the filling. Therefore, a significant effort is taken in order to replace the liquid electrolyte with a gel electrolyte, solid-state electrolyte, or p-conducting polymer material. The best efficiencies obtained so far using p-type conducting materials are in the 1% range. Recently, high efficiencies above 7% were announced by Toshiba using a gel electrolyte.

In terms of a possible integration of these cells into electronic devices, the necessary sintering step of the nanometer-sized TiO_2 at temperatures above 400°C on the transparent conductive oxide (TCO) glass might be a certain drawback. Due to this thermal budget, the glass electrode is the only element that limits the shape of the cells to a flat design.

5.3.3 Organic Solar Cells

Besides dye-sensitized solar cells, which may be considered organic/anorganic hybrid cells, other types of organic solar cells have become of broader interest. These cells can be divided roughly into molecular and polymer organic solar cells [84] or into flat-layer systems and bulk heterojunctions.

Organic materials, e.g., conjugated polymers, dyes, or molecular organic glasses, can show p- or n-type semiconducting properties. Extremely high optical absorption coefficients are possible with these materials, which offers the possibility for the production of very thin solar cells (far below 1 μm), and, therefore, only very small amounts of material are needed. The variability of organic compounds is nearly infinite. Besides this, the large interest in these materials results from technological aspects as the expected ease of large-scale manufacturing at low-temperature processes and very low cost. The upscaling of organic solar cells into large-area devices, always a big challenge with inorganic solar cells, has already been demonstrated to be straightforward. The energetic payback time of organic solar cells is expected to be very short. Considering the fact that light-emitting films of plastic materials have been realized, there is a realistic chance of achieving efficient photovoltaic conversion in such materials, because this is nearly the reverse process. Organic solar cells offer the hope of being very inexpensive. Quite a variety of materials, compositions, and concepts are being investigated, which reflects the possibilities in terms of device concepts, materials use, and materials design. In spite of the many fundamental questions that still exist, these perspectives and the fact that exploration has only just begun cause a greatly growing interest in the development of such solar cells.

The width of the charge generation layer for organic solar cells is much smaller than for inorganic cells. In order to overcome this obstacle, the concept of interpenetrating networks with modified fullerene (i.e., C60) particles was successfully introduced. Fullerenes have been proven to be very efficient electron acceptors for photoexcited conjugated polymers; the quantum efficiency for charge separation is near unity [85]. Mixtures of fullerene derivates (e.g., PCBM) and conjugated polymers (e.g., MDMO-PPV) are spin-coated or doctor-bladed on suitable substrates such as ITO-coated polymer foils or glasses. Interpenetrating networks can also be produced by the coevaporation of fullerenes and molecular dyes such as zinc. Another concept is the use of stratified layers made of donor-acceptor blends, avoiding the problem of continuous and separate pathways for the two types of charge carriers, which in interpenetrating networks might exist between anode and cathode.

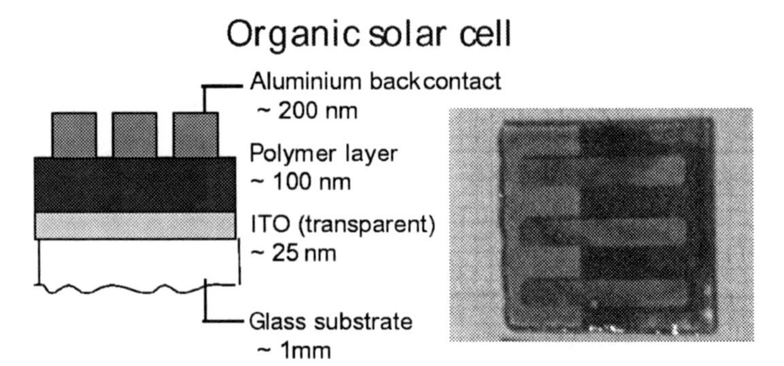

Fig. 5.14. Experimental organic solar cell

Only modest solar conversion efficiencies of up to 1% were reached until 1999. Efficiencies have increased rapidly within recent years: With molecular flat-layer systems based on molecular organic single crystals made of iodine- or bromine-doped pentacene, efficiencies of up to 3.3% under AM 1.5 illumination have been reported at Lucent Technologies. Nearly the same value was reported with improved bulk heterojunctions (an interpenetrating network) of conjugated polymers and fullerene derivatives.

Before these cells become practical, which at the moment still seems far away, the efficiency will have to be further increased. Also, long-term stability and protection against environmental influences are significant challenges. An experiment polymer solar cell is shown in Fig. 5.14.

5.4 Theoretical Concepts for New High Efficiency Semiconductor Materials

This chapter deals with new theoretical materials and concepts that do not yet exist, but would offer significant advantages for photovoltaic conversion. They might be called designer materials. In principle, they are designed to combine the effect of tandem cells in one solar cell.

5.4.1 Auger Generation Material

In this concept, the higher energy photons (energy greater than 2 EG) should generate two or even more electron-hole pairs by impact ionization [86]. The maximum theoretical efficiency of such a material is 42%, instead of 30% for a semiconductor with optimal gap. The requirements for the band structure of such a material have been worked out, but no such material has yet been synthesized. According to [87], the new material should have a fundamental indirect gap of 0.95 eV and a direct gap of 1.9 eV. As a possible material a Si-Ge alloy has been identified.

A related concept is the hot carrier device, in which carriers are collected before thermalization sets in [88].

5.4.2 Intermediate Metallic Band Material and Up and Down Conversion

The intermediate metallic band material solves a problem that has a long history in solar cell materials design. Photons with energy less than the gap could be utilized if an intermediate energy level around midgap were present through which carriers could be transported from one band edge to the other by two photons. Unfortunately, such levels are also strong recombination centers and lead to drastic degradation of the material. This problem is circumvented (at least theoretically) by placing a narrow metallic band within the gap of a wide gap semiconductor [89] (Fig. 5.15). The intermediate band prevents nonradiative recombination. Both holes and electrons exist as minority carriers and, due to the metallic nature of the band, charge neutrality is always established. Optimum absorption edges should be at 0.93, 1.40, and 2.43 eV. The theoretical efficiency of this device is 46.0%, compared to 41.9% of the tandem cell described above. The intermediate band base material should be placed between two ordinary semiconductors, one strongly n-doped, the other strongly p-doped, as shown in Fig. 5.15. The metallic band would otherwise short circuit the device.

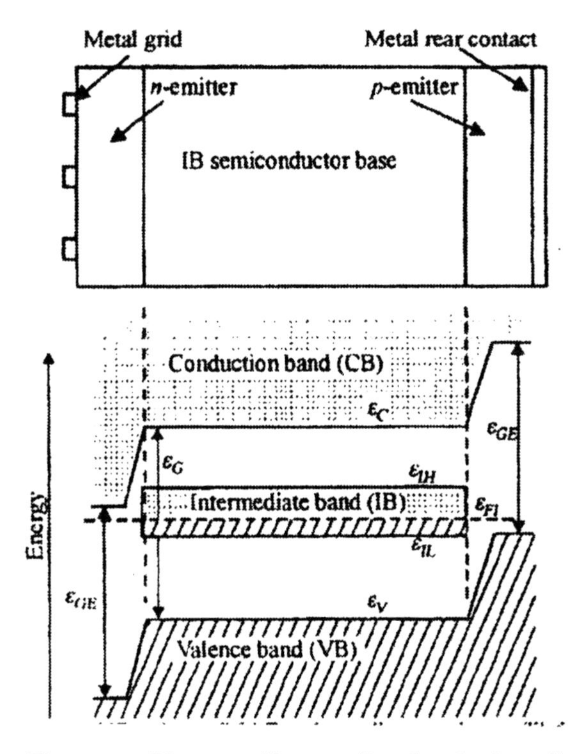

Fig. 5.15. Diagram of intermediate band solar cell

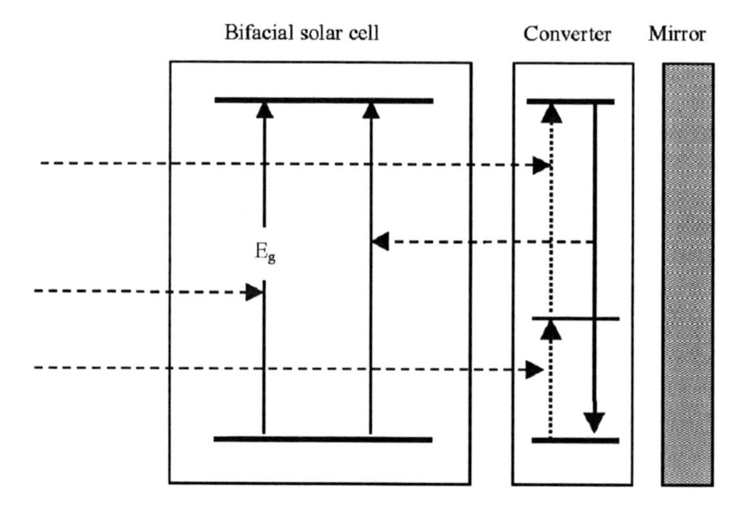

Fig. 5.16. Enhancement of efficiency by adding an up-converter and a mirror to a bifacial solar cell

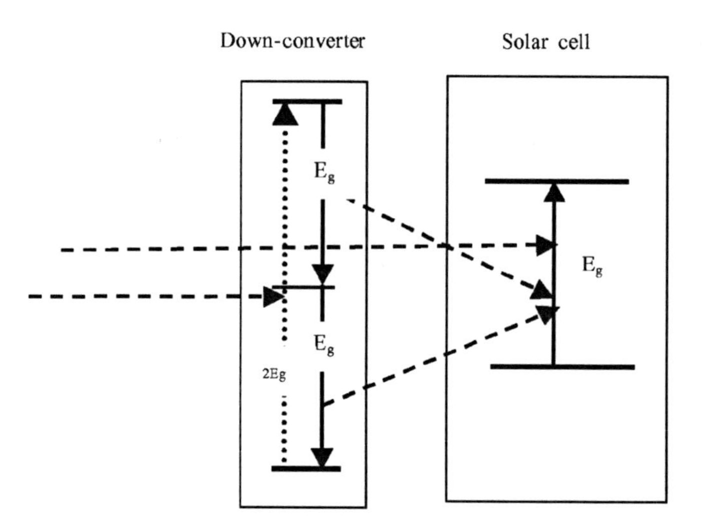

Fig. 5.17. Down conversion of light with energy $E < E_g$

A proposal already exists to manufacture the intermediate band material [90]. By incorporating quantum dots of 35 to 70 Å diameter of a suitable ternary compound into another, wider gap ternary material. Whether this material can be synthesized remains an open question.

A very similar but potentially more practical concept is optical up or down conversion. An up-conversion device is shown in Fig. 5.16. It consists of a bifacial cell followed by an up-converter and a mirror. The up-converter can absorb sub-band gap light not absorbed by the solar cell and is able to stack two quanta to one of useable band gap light. The mirror returns light emitted in the wrong direction to the solar and also reflects sub-band gap light that was not absorbed into the up-converter. This concept as well as the following one has the advantage of separating the electrical and optical functions. Each material can be optimized separately. Bifacial silicon cells, as described above, already exist and even an inefficient up-converter would increase efficiency.

The same principle can be applied to light with more than twice the band gap energy. In this case, high energy quanta are converted into two quanta of band gap energy by a down-converter. Such a material could be placed in front of the cell, as shown in Fig. 5.17, but then there will be losses because some converted light will be emitted to the outside. This loss, however, is less than might be expected, because, depending on the index of refraction of the converter material, light will be trapped by total reflection. If the down-converter is placed behind the cell, the cell material has to fulfill stringent requirements regarding light absorption. It has to be transparent to higher energy quanta, which means that valence and conduction band have to be very narrow.

5.5 Past and Future Development of Solar Cell Efficiency

The future development of crystalline silicon and the other materials will now be described by a method that was introduced recently [91]. We can define a function that models the past development of solar cell efficiency and also permits prediction of future development. The basic requirements for this function are relatively easy to define: It should start at some point in time and should have a saturation value that it will reach after a very long time of development. The function chosen here is relatively simple (it has the same structure as the diode equation): $\eta(t) = \eta_{\mathrm{L}}\big(1 - \exp((a_0 - a)/c)\big)$.

Figure 5.18 shows this function and its application to the maximum laboratory values of silicon solar cells. The experimental values are shown together with the fit function. It has three adjustable parameters that are all determined empirically by fitting. Most interesting besides the good fit is the final efficiency, which comes out to be 29%, but it can only be reached after infinite time. It can also be seen that improvement of efficiency from now on will be very slow.

We can also fit this function to production modules. Here values are much more difficult to collect, and the curve in Fig. 5.19 is based on very few points. In this graph, module efficiencies, not cell efficiencies in modules, were used. This led to the result that the highest efficiency module was a polycrystalline module, because this technology has a better area utilization. From this graph, we extrapolate a limiting efficiency of 21%. Another conclusion

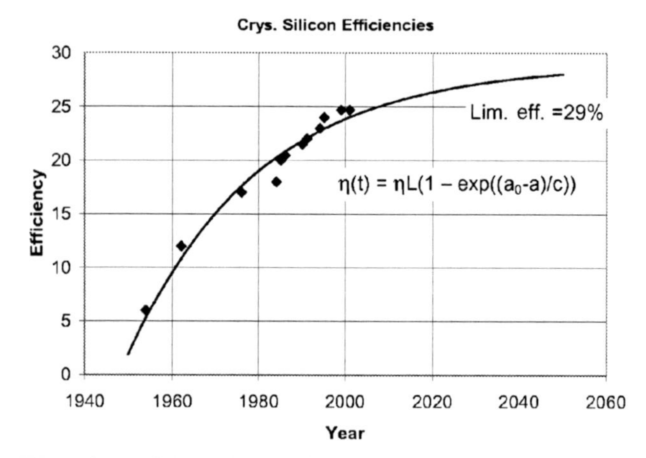

Fig. 5.18. Time dependence of crystalline silicon solar cell efficiency. *Points* are actual values of maximum efficiency and *solid line* is fitting function

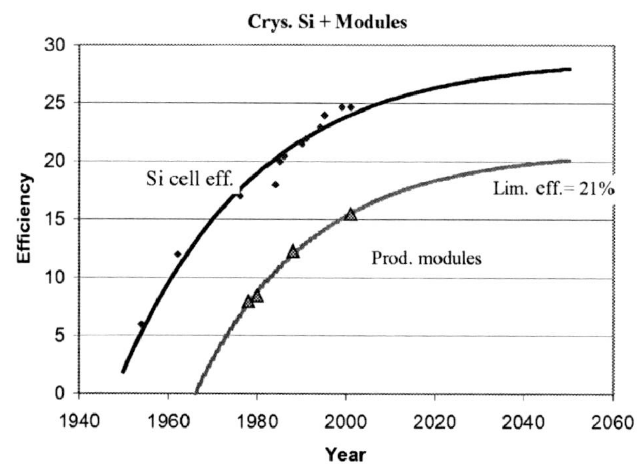

Fig. 5.19. Maximum production efficiency together with maximum laboratory values

can be drawn from these curves: It is normally assumed that production efficiency is a certain number of years behind the laboratory values. In reality this gap increases! This is actually logical, because if the limiting efficiency of production modules is lower, the time difference between the two curves will become infinite.

Next, thin-film crystalline silicon efficiency is presented in its historic and future developments. The highest efficiencies are determined by the transfer technique described in Chap. 4. In Fig. 5.20, the efficiency development of thin-film crystalline silicon is shown with that of bulk silicon for comparison.

Compared to bulk silicon the Si thin-film technology shows very rapid growth. (Nevertheless, it will not reach the present bulk Si efficiency until

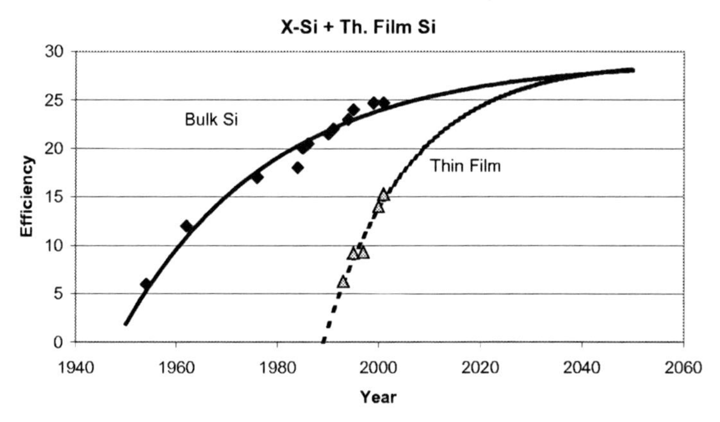

Fig. 5.20. Development of thin-film crystalline silicon

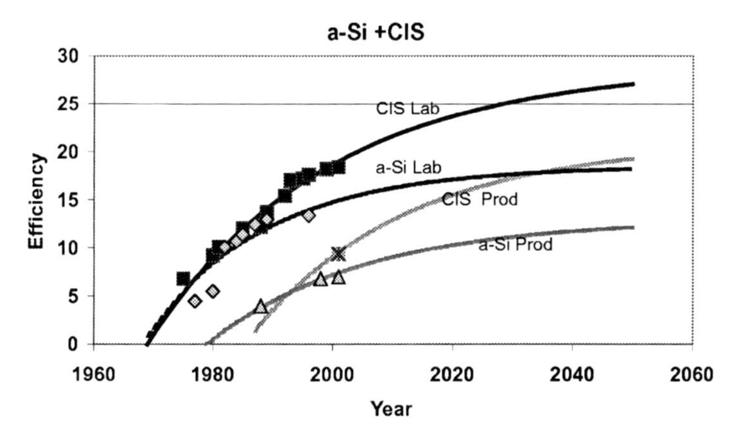

Fig. 5.21. Laboratory and production efficiency for CIS and a-Si

2020.) The limiting efficiency was set equal to bulk silicon. It is still open as to when and at which cost those results can be transferred into production.

A similar plot can be drawn for the so-called true thin-film materials a-Si and CIS. (Fig. 5.21). The following conclusions can be drawn from this plot: CIS has the potential to reach the limiting efficiency of crystalline silicon. It has a much higher limit than amorphous silicon.

The data for a-Si have a large scatter and, therefore, the conclusions have to be taken with care. The limiting laboratory efficiency for a-Si is about 18%. For production this value is 13%. The production curve for CIS is based on only one point while the other parameters are taken from standard silicon. This curve follows the curve for silicon with a delay of about 20 years.

An overall perspective of most of the present and two potential future technologies is shown in Fig. 5.22. Two hypothetical curves are included in this graph: Organic cells are based on only one point, the other parameters are estimated and a hypothetical new high-efficiency third-generation material or

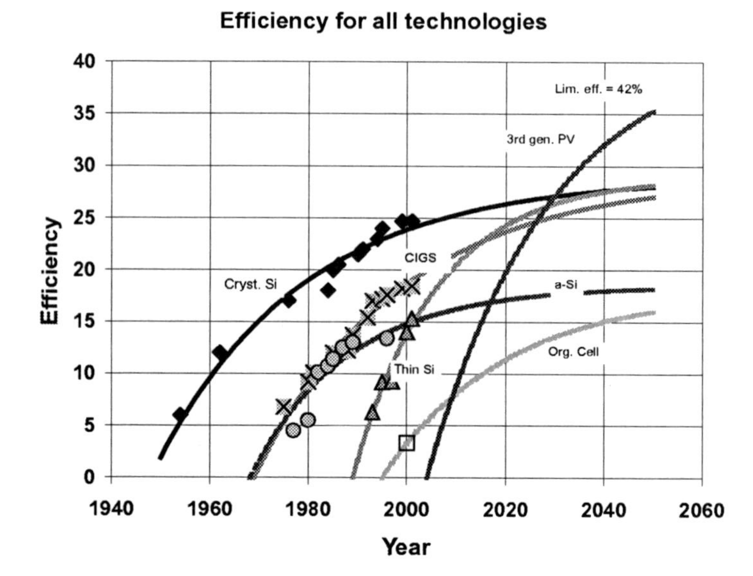

Fig. 5.22. A century of photovoltaics. Past achievements and future predictions of highest laboratory solar cell efficiencies. The best data are for crystalline silicon. The right curves are based on very few data points, the curve for a new material is purely hypothetical

Table 5.7. Fitting parameters for different technologies

Technology	η_L	c	a_0
Xtal Si	29	30	1948
Thin-film Si	30	19	1989
CIS/CIGS	29	30	1969
a-Si	18	20	1968
Organic cells	18	25	1995
New material	42	25	2004

combination of materials is shown (curve to the right), the development of which is (optimistically) assumed to be starting now. All fitting parameters of the curves are given in Table 5.7. For the curves in Fig. 5.22 to become a reality a precondition is that somebody is working on those developments. This is unlikely in the asymptotic part of the curves, because a large effort is required for a very small improvement.

6 Solar Cells and Solar Modules

6.1 Characteristic Curves and Characteristics of Solar Cells

The solar cell is an electric component with some properties that differ markedly from those of "customary" energy sources. Knowledge of the characteristics of these solar cells is however a prerequisite for designing and dimensioning a photovoltaic power supply for stand-alone appliances or grid-connected systems, for achieving reliable installation, and, in particular, for detecting errors and commissioning photovoltaic systems. Furthermore, it allows the possibilities and limits of a photovoltaic power supply to be recognized and thus ensures that information to prospective customers and users is well-founded.

In the next chapters, the essential characteristics of silicon solar cells of solar modules will be presented. Most of the information was adapted from [92].

6.1.1 Characteristic Curves of Solar Cells

The cross section through a crystalline solar cell given in Fig. 1.1 shows that it is in principle a large-area silicon diode. In the dark state, the characteristic curve of this diode essentially corresponds to the well-known curve of a normal diode (Fig. 6.1). (See also Fig. 2.7 in Chap. 2, which only shows the first quadrant of the characteristic.)

In the forward direction (quadrant I), practically no current flows initially at low voltages, but above a voltage of ∅ 0.4–0.6 V it increases rapidly. In the reverse direction (quadrant III), current flow is blocked up to a certain voltage limit (for solar cells, a few to tens of volts), at which the diode "breaks down," i.e., becomes conductive – generally this results in the destruction of the component. By contrast, an externally induced current many times larger than the rated current in the forward direction does not cause problems.

As already outlined in Chap. 2, illumination of the solar cell creates free charge carriers, which allow current to flow through a connected load. The number of free charge carriers created is proportional to the incident radiation intensity, so that the photocurrent (I_L) internally generated in the solar cell

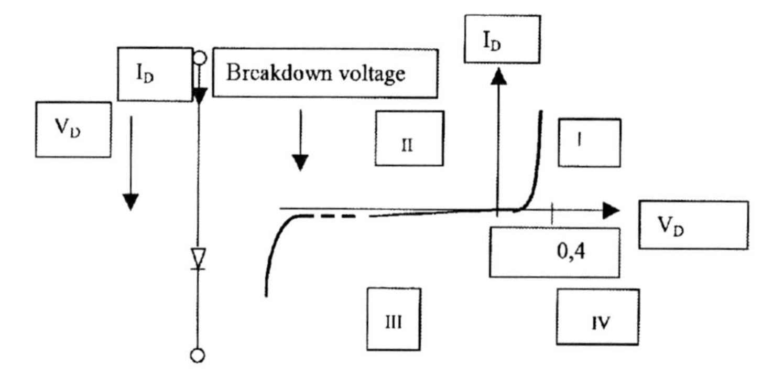

Fig. 6.1. Circuit diagram symbol and current-voltage characteristic for a silicon diode. (The subscript D refers to Diode)

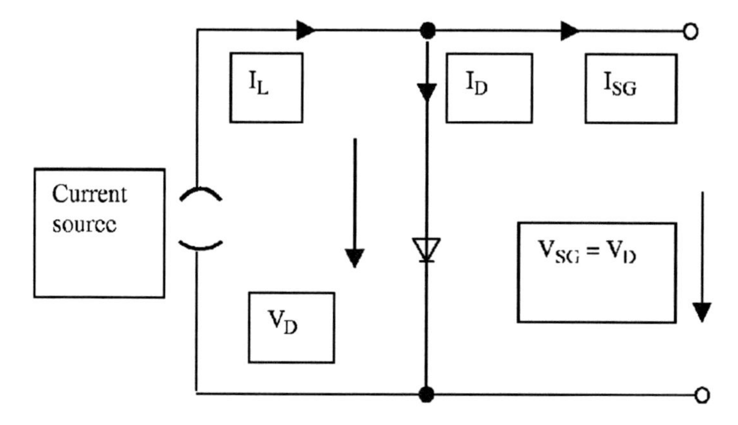

Fig. 6.2. Simplified equivalent circuit diagram for a solar cell. (The subscript SG refers to Solar Generator)

is also proportional to the radiation intensity. Thus, the simplified equivalent circuit diagram shown in Fig. 6.2 can be used to represent a solar cell.

It consists of the diode created by the p-n junction and a current source, with the magnitude of the current depending on the radiation density. As shown in Chap. 2, a more accurate physical model works with two diodes with different physical properties (Fig. 2.8).

6.1.2 Characteristics of Solar Cells

In the following, the quantities needed to characterize both solar cells and solar modules will be explained using crystalline solar cells as an example, but the concept is also applicable to other types of solar cells.

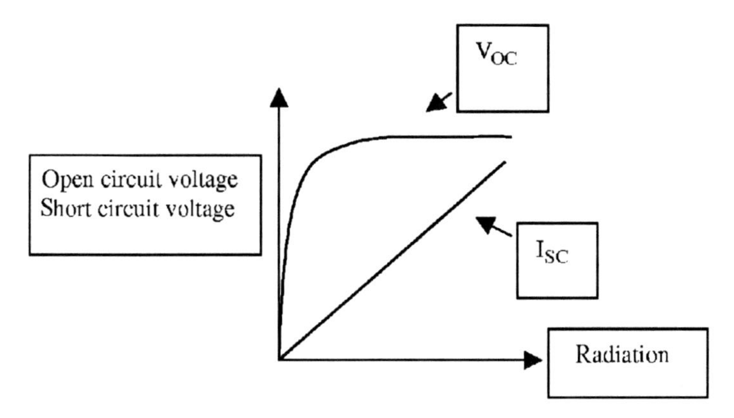

Fig. 6.3. Open circuit voltage and short circuit current as a function of the radiation intensity

Short Circuit Current

As mentioned above, the short circuit current I_{SC} is proportional to solar radiation over a wide range (Fig. 6.3). Also, the short circuit current depends on cell temperature. It increases for crystalline silicon solar cells by about 0.05–0.07%/Kelvin and for amorphous silicon modules by about 0.02%/Kelvin (1 Kelvin corresponds to a temperature difference of 1°C). The lower temperature coefficient of amorphous modules is one of the reasons why they perform better than crystalline modules in hot summer months.

Open Circuit Voltage

The open circuit voltage or V_{OC} corresponds to the voltage across the internal diode, when the total generated photocurrent flows through this diode. According to Fig. 2.4, the voltage increases very rapidly with illumination until it reaches a near saturation value. At this point, it increases very slowly as $\ln(I_L/I_O)$. Very often this increase is not seen because of internal and external resistances (Fig. 6.3). Crystalline silicon solar cells reach a typical value of 0.5–0.6 V and amorphous silicon cells reach 0.6–0.9 V.

Whereas the open circuit voltage and also the working point voltage can be assumed to be almost independent of the radiation value for the typically high intensities outdoors, these voltages drop markedly in poorly lit indoor rooms with intensities of only a few W/m^2.

Furthermore, the open circuit voltage, and also the working point voltage, are strongly dependent on temperature (see “Maximum Power Point (MPP)” in this section). This must be considered, as solar cells installed outdoors can reach temperatures, depending on the installation (e.g., possibilities for ventilation), up to 40 K higher than the ambient temperature (see “Short Circuit Current” in this section also).

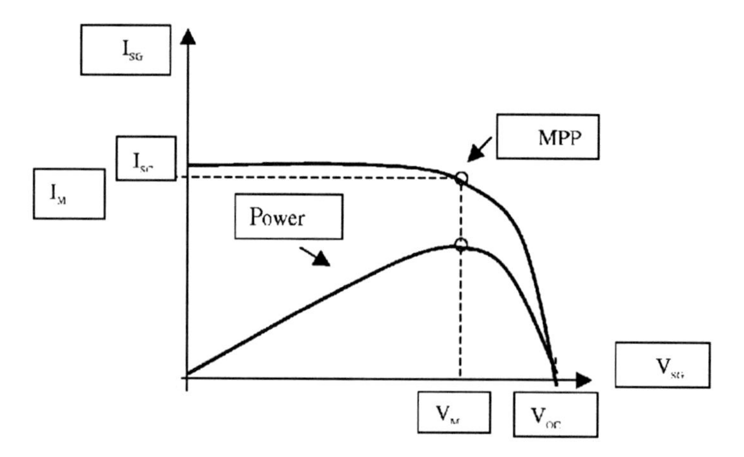

Fig. 6.4. Power curve and maximum power point (MPP)

Power

The power delivered by a solar cell is the product of current and voltage. If the multiplication is done, point for point, for all voltages from short-circuit to open-circuit conditions, the power curve illustrated in Fig. 6.4 is obtained for a given radiation level.

Although the current has its maximum at the short circuit point, the voltage is zero and the power is also zero. The situation for current and voltage is reversed at the open circuit point, so again the power here is zero. In between, there is one particular combination of current and voltage for which the power reaches a maximum (graphically the area of the rectangle indicated). This is the maximum power point or MPP (see next paragraph).

Maximum Power Point (MPP)

The MPP of the solar cell (respectively, the solar module or the solar generator) is positioned near the bend in the curve shown in Fig. 6.4. The corresponding values of V_M and I_M can be estimated from the open circuit voltage and the short circuit current:

$V_M \sim (0.75\text{–}0.9)\, V_{OC}$
$I_M \sim (0.85\text{–}0.95)\, I_{SC}$.

Because cell voltage and current depend on temperature (see "Short Circuit Current" and "Open Circuit Voltage" in this section) the supplied power also changes with temperature. The power of crystalline silicon solar cells drops by about 0.4–0.5%/K. and the power of amorphous silicon solar modules drops by about 0.2–0.25%/K.

The rated power of a solar cell or a solar module is measured under internationally specified test conditions (STC = Standard Test Conditions) with the following parameters:

- Radiation intensity — 1,000 W/m^2
- Temperature of the solar cell — 25°C
- Air Mass — 1.5[1]

The rated power is reported in Wp (peak watts). It should be noted that the most commonly used term "peak power" is misleading, because at lower cell temperature or higher radiation intensities the reported value can well be exceeded. Mostly, however, the modules operate at much lower radiation.

Efficiency Value

Only part of the solar radiation incident on the solar cell is converted to electricity. The ratio of the output electrical energy to the input solar radiation is defined as the efficiency value. It depends on the type of cell (see Chap. 2).

For the module efficiency value, the output power is divided by the total radiation incident on the module. Because the entire area of the module is not covered with solar cells (for example, frames and space between the single solar cells are not active area), the module efficiency value is lower than the efficiency value of the single cell.

Spectral Sensitivity

Depending on the technology and the material used, solar cells vary in their sensitivity to the different spectral ranges of the incident radiation. As an example, Fig. 6.5 illustrates the relative spectral sensitivity (or spectral response) of an amorphous silicon solar module and a crystalline silicon solar cell. It can be seen that the latter has a higher sensitivity in the long-wavelength range, whereas the amorphous silicon module is most sensitive in the visible spectral range. As was shown in Chap. 5, absorption of different wavelengths depends on the band gap of the solar cell material.

Fill Factor

The simplified equivalent circuit diagram, shown in Fig. 6.2 to explain the fundamental behavior of crystalline silicon solar cells, must be modified to include two resistors for a more exact description. The series resistor R_S is composed of the resistance through the silicon wafer, the resistance of the back surface contact and the contact grid on the front surface, and, further, the circuit resistance from connections and terminals. The parallel (or shunt) resistor R_P results, in particular, from the loss currents at the edges of the solar cell and surface inhomogeneities (Fig. 6.6).

---

[1] Definition of Air Mass (AM): The Air Mass number refers to the intensity and spectral distribution resulting from a certain path length of sunlight through the atmosphere. AM 1 means the sun is directly overhead. AM 1.5 applies at a sun position such that the path length is 1.5 times more.

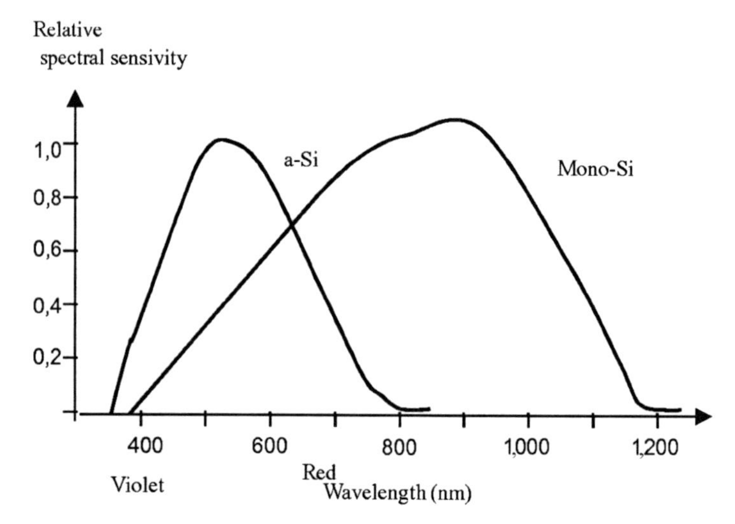

Fig. 6.5. Relative spectral sensitivity (spectral response) of an amorphous silicon solar module and a crystalline solar cell

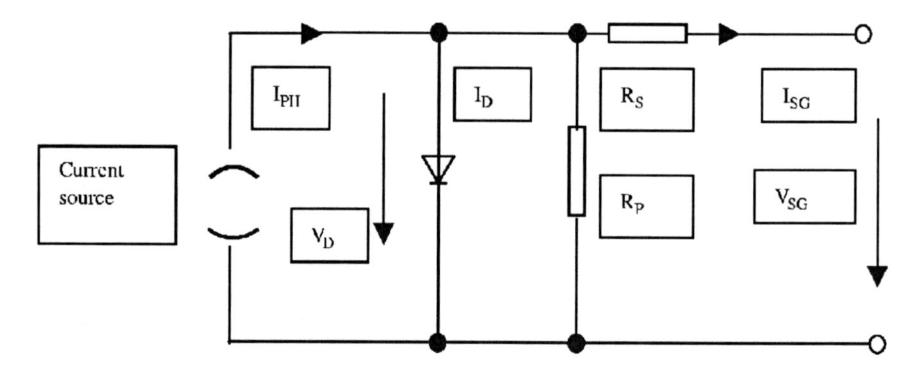

Fig. 6.6. Extended equivalent circuit diagram for a crystalline silicon solar cell. The subscript SG refers to Solar Generator

Both resistors make the characteristic curve less rectangular, and the maximum power output is reduced. A further measure for the quality of a crystalline silicon solar cell is therefore the fill factor (FF), which describes how closely the current-voltage characteristic curve approximates the ideal rectangle form, see Fig. 2.5.

The fill factor for crystalline silicon solar cells is about 0.7 to 0.8, and the fill factor of amorphous silicon solar modules is about 0.5 to 0.7.

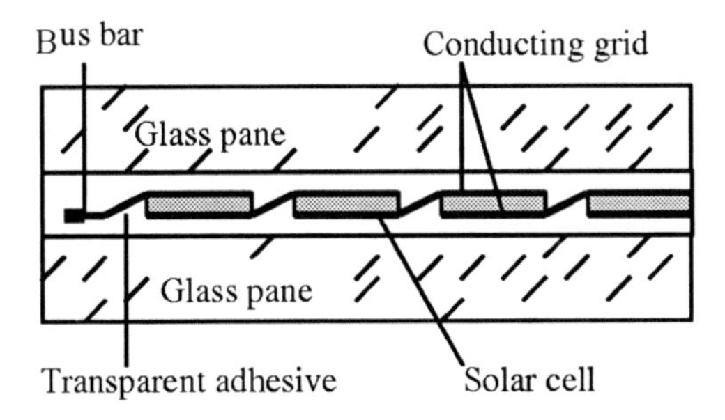

Fig. 6.7. Cross section through a typical module of crystalline silicon solar cells

6.2 Module Technologies

One single crystalline silicon solar cell with a surface area of approximately 100 cm^2 generates a current of 3 A at a voltage of 0.5 V when exposed to full sunshine. Up to five years ago, the typical PV module made of crystalline silicon consisted of 30 to 36 cells connected in series with a peak power of approximately 50 W. Today, modules with a peak power up to 300 W are being marketed. Such a module consists of more than 100 solar cells connected in series and parallel. More information about the connection principles of solar cells and solar modules can be found in Chap. 7.

In the case of thin-film materials (e.g., amorphous silicon or copper indium diselenide), complete modules are manufactured. So the above mentioned step from single solar cell to a solar module is not necessary.

A cross section through a typical module of crystalline silicon solar cells is shown in Fig. 6.7. The module's top layers are transparent. The outermost layer, the cover glass, protects the remaining structure from the environment. It keeps out water, water vapor, and gaseous pollutants thath could cause corrosion of a cell if allowed to penetrate the module during its long outdoor use. The cover glass is often hardened (tempered) to protect the cell from hail or wind damage. A transparent adhesive holds the glass to the cell. The cell itself is usually covered with an antireflective coating. Some manufacturers etch or texture the cell surface to further reduce the reflection, as described in Sect. 3.4.

The cell's bottom layer is called the back contact and is a metal film, which in connection with the front contact forms a bridge to an external circuit. The module's back side is either covered with a layer of Tedlar™ or glass, as shown in Fig. 6.7. Often a frame of aluminum or composite material gives the module the needed mechanical stability for mounting it in different ways (see Chap. 8).

A single crystalline silicon solar cell generates electric power in a range of 1.5 Wp at the maximum only. In most practical cases, this is not enough. Therefore, it is necessary to interconnect a certain number of solar cells to a solar module. Depending on the power range of a PV module, the connection of the separate solar cells can be realized in series only or both in series and parallel.

In industrial manufacturing, the connection of solar cells to produce a PV solar module requires a number of different steps, which will not be described in detail here. It makes a great difference whether the PV modules are produced in a plant with great output, maybe 20 MWp per year or in a smaller plant with an output of less than 1 MWp per year. The production process in a plant with a high yearly output of PV modules is fully or nearly fully automated. In the small PV solar module plants, many production steps are not automated. They are carried out by hand or with simple mechanical devices.

The degree of automation of the production process has great influence on the production cost of the PV solar modules (see Chap. 11). Today, PV solar modules have a power of up to 300 Wp. Because commercially used PV systems mostly need a higher power than can be delivered by a single PV module, the connection of several PV modules to a so-called solar generator is necessary.

The same connection principles as for the connection of solar cells apply to modules, i.e., they can be connected in series and parallel or in series only.

Series Connection

Figure 6.8 shows a series connection of solar cells. Here, the same current flows through every solar cell, and the total voltage is the sum of the partial voltages across the individual cells.

Series connection of the solar cells and also of the solar modules causes an undesired effect when a solar cell or module is fully or partly shaded. The weakest link in the chain determines the quality of the whole system. Even when only one cell is (partly) shaded, the effect is the same as if all the series-connected cells or modules were shaded. In this way the power output drops drastically. Thus, it is imperative to avoid even slight shadows, e.g., from

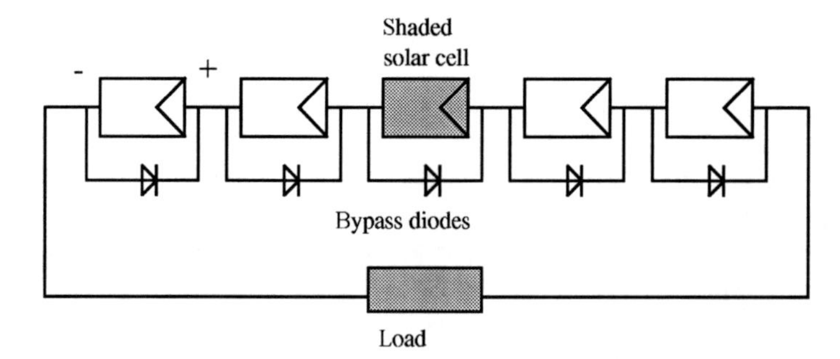

Fig. 6.8. Series connection of solar cells and solar modules. The bypass diodes prevent the occurrence of hot spots when one of the cells is shaded

cables, mounting wires, or tree tops. A further intolerable effect caused by series connection is the occurrence of local hot spots when individual cells are partly shaded. The occurrence of the hot spot can be understood by realizing that the shaded diode presents a very high resistance compared to the load. Then most of the voltage drop generated by the rest of the cells appears at the shaded diode, which is driven into breakdown.

To avoid this operating condition, so-called bypass diodes are connected antiparallel to the solar cells such that larger voltage differences cannot arise in the reverse-current direction of the solar cells. The ideal solution is shown in Fig. 6.8, with one such bypass diode for each solar cell. However, in practice, it is sufficient to provide one bypass diode for every 15 to 20 solar cells. In general, connections for these are included by the manufacturer in the connection box.

Because, as mentioned above, when connecting solar modules in series the weakest link in the chain determines the quality of the complete string solar modules of different technologies or from different manufacturers should not be series connected to a solar generator. This also applies for the series connection of several solar cells to a module. In practice, differences in the efficiency of solar cells or modules between the individual production charges from the same manufacturer are possible. This situation is called mismatch, and the losses are called mismatch losses.

Parallel Connection

If higher currents are demanded in a system, these can be obtained by parallel connection of the individual strings, as is shown in Fig. 6.9. In a parallel-connected configuration, the voltage across each solar cell or solar module is equal, while the total current is the sum of all cell or module partial currents.

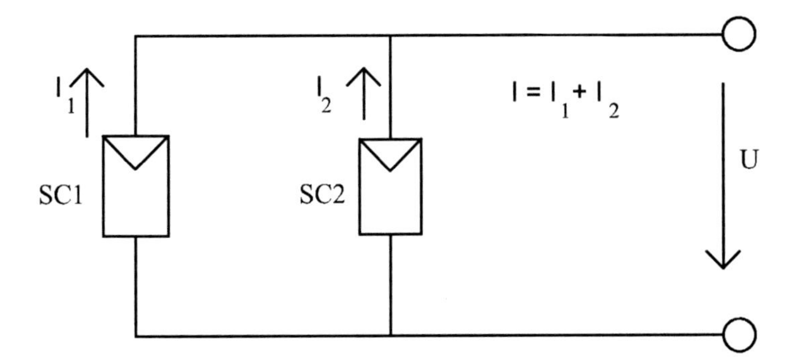

Fig. 6.9. Parallel connection of solar cells and solar modules

Orientation and Elevation of Modules

It is evident that a solar module will deliver maximum power if it is oriented perpendicular to the incoming radiation. This can only be achieved in tracking systems, but in most applications the modules are installed in a fixed position. Then the best time averaged irradiation has to be chosen as a compromise. For generators in Central Europe the following boundary conditions are recommended: Orientation toward the south, but a deviation of 10 to 15 degrees does not appreciably reduce output. Inclination should be toward the horizontal 30°–42° for an optimized output over the full year. If output is to be optimized for spring, fall, and winter, the inclination should be increased to 45°–60°. Other geographic locations require different optimization rules. In general, in all locations in the northern hemisphere, a solar generator should be oriented as much as possible toward the south and in the southern hemisphere toward the north. For the inclination in dependence on geographic latitude β, the following rule for optimized yearly yield applies:

$$\beta = \text{geographical latitude} \pm 10^\circ \,. \tag{6.1}$$

7 PV Systems

Terrestrial photovoltaic applications can be divided into:

- stand-alone PV systems and
- grid-connected PV systems.

Figure 7.1 shows the annual percentage of grid-connected PV systems and off-grid or stand-alone PV systems on the world PV market from 1990 to 2002. Clearly, it can be seen that the percentage of the grid-connected increased rapidly during this time [93].

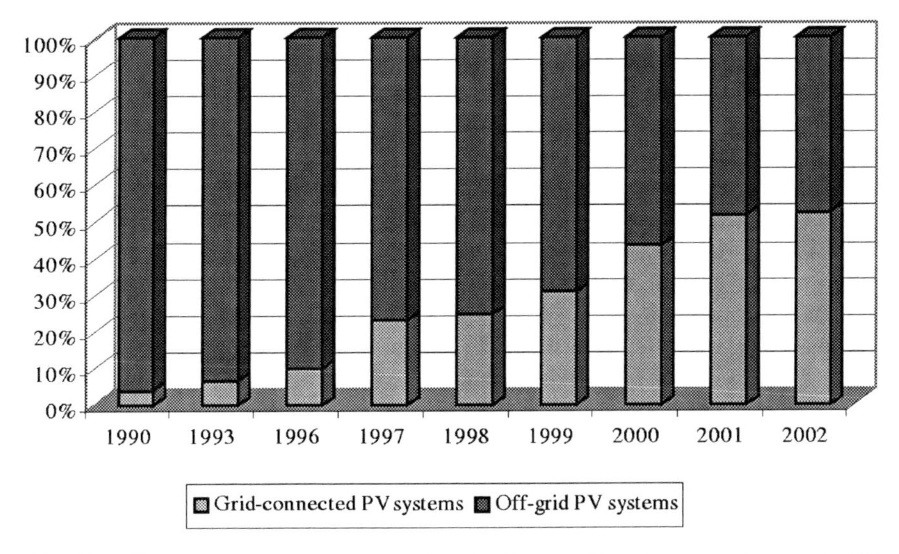

Fig. 7.1. Percentage of grid-connected and off-grid PV systems on the world PV market [93]

7.1 Stand-Alone PV Systems

Stand-alone photovoltaically powered systems with peak PV powers can have from milliwatts to several kilowatts. They do not have a connection to an electricity grid. In order to ensure the supply of the stand-alone system with electric power also in the times without radiation (e.g., at night) or with very

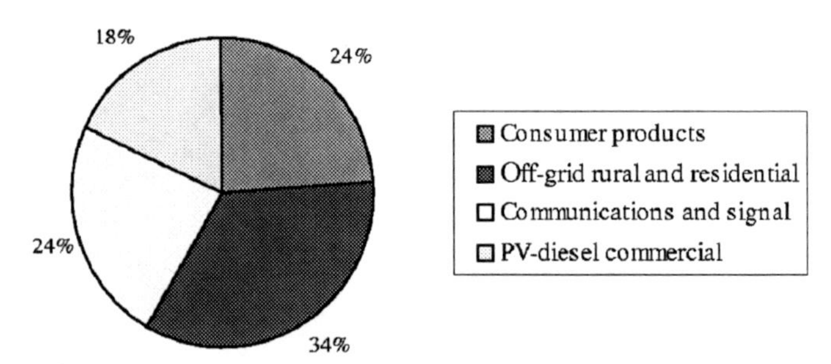

Fig. 7.2. Percentage by application area on the off-grid or stand-alone PV world market in 2002 [93]

low radiation (e.g., at times with a strong cloud cover), stand-alone systems mostly have an integrated storage system. If the systems are used only during the time when the radiation is sufficient to supply the system with electric power directly, a storage system is not necessary. This also applies to the situation in which the product delivered by the system can be stored (e.g., water).

At present, a very great variety of stand-alone system exist. Examples range from solar calculators and watches to systems for traffic control to systems that are able to supply one or several buildings in remote areas with electric power. They can be dc systems with or without a storage battery, or they can be ac systems with an inverter.

Stand-alone systems can be implemented with a PV generator as the only power source or with auxiliary power sources, as so-called hybrid systems (see Sect. 7.1.4), where additional generators employing fossil fuel (for example, diesel fuel or gas) or renewable energy (for example, wind, hydropower, or biomass) complement the PV energy production. The choice of storage capacity and of the relative power of a PV generator and an auxiliary power source depends on radiation conditions, the required security of supply, and last but not least on economics. Today, system designers can use layout programs that make it easy to find the optimum.

Figure 7.2 gives an overview of the percentage by application on the offgrid respectively on the stand-alone PV world market. More than a third of all stand-alone PV systems are used for rural and residential off-grid systems.

7.1.1 Consumer Applications

Solar calculators were among the first consumer applications. This is an ideal application because no storage is needed. The solar calculator works whenever there is enough light to read the display. Today, solar watches, battery chargers, and various other products are available with solar power. All the products are equipped with small integrated PV modules, a voltage DC/DC

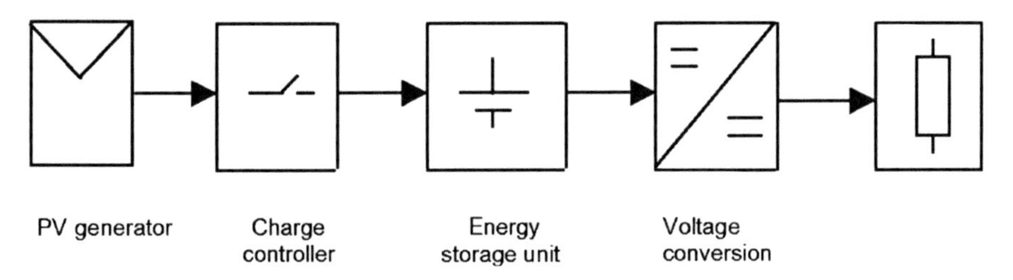

Fig. 7.3. Block diagram of a photovoltaic power supply for a small consumer product

Fig. 7.4. Two examples of photovoltaic appliances

converter, and, in most cases, an energy storage unit and a charge controller (Fig. 7.3).

In Fig. 7.4, some examples of photovoltaic appliances are presented. From left to right: a PV system to meet the electricity demand of a telephone on a railway line and an organizer with an integrated small PV module.

7.1.2 Solar Home Systems

According to the estimates of the European Union, about two thousand million people throughout the world do not have a connection to a public grid. And about half of these live in regions without any access to electricity. High investment costs in combination with a low electricity consumption (less than 1 kWh per day) will act against the grid being extended to these remote, sparsely populated regions within the foreseeable future.

For these regions, the supply with electricity can only be realized by decentralized, small power plants. Mostly diesel generators were used. But their operation is connected with adverse environmental effects like noise and noxious gases. In addition, an infrastructure for supplying the diesel generators with fuel oil and maintenance must exist in outlying locations. However, in

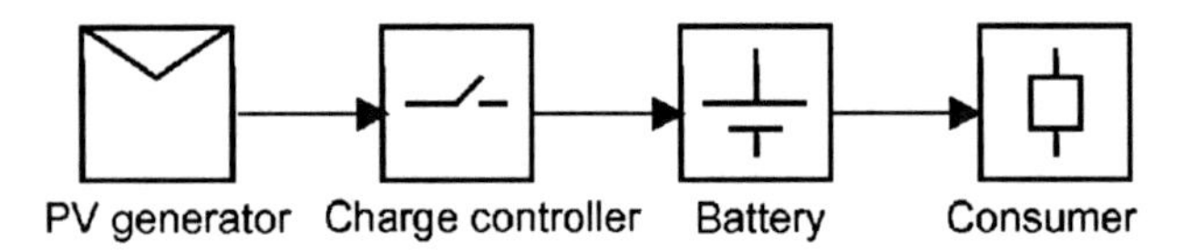

Fig. 7.5. Schematic principle of a simple solar home system

the remote and sparsely populated regions in Africa, Asia, and Latin America there are rarely all-weather streets or rail connections. It is not surprising, therefore, that only very few households can afford such generators. Photovoltaics in the form of so-called solar home systems is seen as the best candidate to solve the energy supply problems in these territories.

A simple Solar Home System (SHS) consists of the solar generator (PV module), lead battery, and charge controller, as well as the directly connected DC appliances (see Fig. 7.5). In addition, a support structure for the module, cables, and sockets for the appliances is needed.

Such an SHS is able to supply the electrical power needed for lighting, TV, radio, or for a small refrigerator.

In a simple SHS, mainly DC-powered electrical appliances are used. As the acceptance of a photovoltaic system hinges on faultless operation of all system components, high quality is demanded for these appliances. In particular, the following are needed:

- long lifetime,
- high efficiency,
- high reliability, even when operated incorrectly,
- correct operation even at extreme temperatures,
- flawless operation for all possible input voltages offered by battery operation, and
- low maintenance and service requirements.

At this time several, hundred thousand SHS are in operation in Africa, Asia, and Latin America (Fig. 7.6).

It is estimated that about 200,000 solar home systems (SHS) are sold annually. And a steady growth is expected for the next years.

Compared to the traditional energy sources used for lighting, i.e., open fire, candles, and kerosene lamps, the benefits of SHS are manifold [94]:

- the improved quality of illumination provides opportunities for extending the work day and for reading and leisure activities;
- there is no health risk associated with smoke and smell; and
- there is no risk of fire hazard.

This improvement in living conditions can lead to a decline in rural exodus.

Apart from this, considering the long lifetime of a SHS – the panel lasts more than twenty years, monthly expenses for traditional energy sources often

Fig. 7.6. Solar home system in Bolivia

exceed the monthly costs of a photovoltaic system. And still, even though the benefits are numerous and a very high technological standard has been achieved, problems have to be faced regarding a widespread dissemination of SHS.

One of the main barriers to the development of rural electrification markets with photovoltaics is the financing of the high up-front investment of an SHS, which is the critical point for most of the rural households, as well as for many photovoltaic intermediaries. Depending on the size of the local market, taxes, the share of locally manufactured components, and governmental policy regarding solar technology and rural electrification, an SHS costs between US $500 to US $1,500. Even with the existence of subsidies for the installation of an SHS, such investment costs are much too high compared to the average income of the target households. For this reason, financing schemes allowing payment by installments must be offered – they are needed in national electrification programs as in commercial markets.

In this connection, micro-finance institutions (MFI) could play an essential role in the dissemination of SHS. This not only represents a solution for the existing barriers regarding the electrification of remote rural households, but also offers new market opportunities for MFI, for example, in form of rural electricity loans, which additionally offer the convenience of secure guarantees of the hardware, especially the solar panel.

To realize this business, the MFI, responsible for the financial flow, could enter into a business relationship with a rural electrification company responsible for the technical flow, i.e., installation, operation and maintenance, and user training. The contract between these entities comprises arrangements on re-sale and repurchase of the solar panel in the case of non-payment of the credit to facilitate using the hardware as guarantee. A part of this agreement on the direct channeling of the credit to the company – after the installation of the system – should be considered to lower the risk, that the client uses the credit for a different purpose [94].

The above mentioned possibility of the combination of micro-financing with the quality of the delivered SHS and user training is very important. The dissemination of SHS can be influenced negatively if the SHS or its components do not work without trouble; i.e., the charge controller often has a breakdown, the battery has a low efficiency, or the efficiency of the solar modules is lower in reality than noted in the data sheet. One reason for this problem is that local companies in some cases sell cheap components of bad quality. As a consequence, the SHS has a low efficiency, a low energy output, and the owner of the SHS is dissatisfied.

Also important is the training of the potential users of an SHS. They should have knowledge of the basic working principle of an SHS. However, most SHS are sold in the rural areas of Africa, Asia and Latin America, where the people are not able to read a detailed instruction manual. So the information about the SHS must be given in a very simple form, i.e., with pictures or as a comic strip. Only if the potential users of SHS are informed about the workings of an SHS or about the workings of the main components of an SHS incorrect operation of the system or incorrect respectively illegal operation can be prevented. An example that is frequently observed is the following: The owner of an SHS has bought additional electric appliances. Now, the SHS cannot meet the increased energy demand when all appliances are operated at the same time. In this case, the SHS owner is tempted to tamper with the charge controller to fully discharge the battery. This works for a short time but seriously decreases battery lifetime.

7.1.3 Residential Systems

Another use of stand-alone PV systems is supplying single buildings with electric power. Whenever houses have no connection to the public grid, photovoltaics is a very attractive economic alternative to a diesel engine.

Even in Central Europe, which has an extended public power grid, there are a few hundred buildings that are not connected to this grid. Most of them are situated in remote areas like mountaineering lodges and other buildings in European mountains (e.g., the Alps and the Black Forest). Currently, in the German part of the Alps more than thirty mountaineering lodges or hikers' inns are supplied with electric energy by stand-alone PV systems with different power ranges (Fig. 7.7).

Fig. 7.7. A PV system supplies a mountain lodge in the German Alps

Mostly the internal grid of the house works with AC power. Therefore, the PV system includes one or two inverters (for the function of inverters, see Sect. 7.2.3). Because the radiation intensity changes with the time of day as well as the season and weather conditions, a stand-alone PV system that is used to supply a building with electrical energy must also have a storage battery and a charge controller. The schematic principle of stand-alone PV system supplying a building is given in Fig. 7.8.

Batteries – In most cases, lead-acid batteries are used for this purpose. Recently, some firms began offering so-called solar batteries specifically designed for stand-alone PV systems. It should be stressed that batteries in a PV system operate in a very demanding environment, in particular, seasonal periods with very low charge are encountered. This situation is alleviated in hybrid systems in which the auxiliary generator can charge the battery during such periods.

Charge controller – A PV battery charge controller serves generally to protect the battery against overcharging and deep discharge. It is absolutely necessary for the efficient operating conditions of the battery and of the complete PV system.

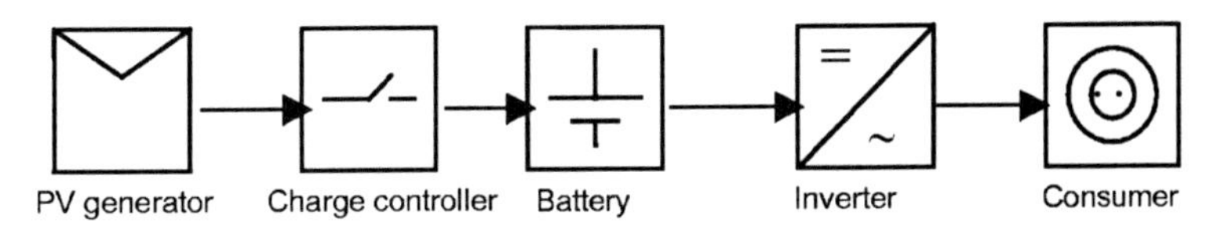

Fig. 7.8. Schematic principle of a stand-alone PV system supplying a building

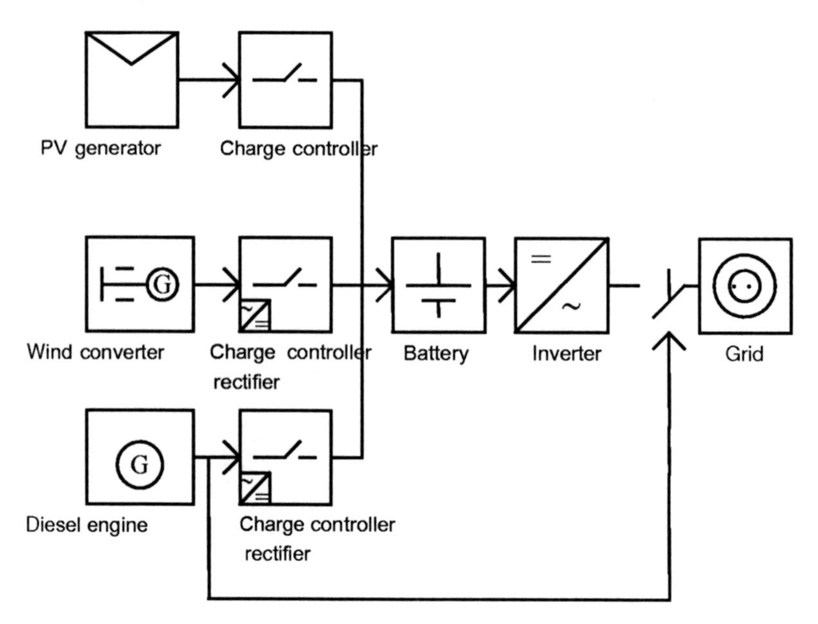

Fig. 7.9. Schematic principle of a hybrid system with PV, wind, and diesel generators

7.1.4 Hybrid Systems

Because of the annual fluctuation of the solar radiation in most parts of the world, an exclusively photovoltaic power supply system would demand a large solar generator and/or a large storage battery. Such a PV system is very expensive. This also applies to photovoltaic systems for which security of supply is very important. Therefore, usually different types of electricity generators are combined into a so-called hybrid system. A combination of photovoltaic and fossil-fuelled generators guarantees the same supply reliability as the public grid.

Under favorable weather conditions, the total energy demand is met by the solar generator. Surplus energy is stored in batteries. During the night or unfavorable weather, the energy demand is initially met by the batteries. If there is danger of deep discharge, a diesel or gas-fuelled generator provides the electricity and simultaneously charges the battery.

At windy sites, the system can include a wind energy converter. As the photovoltaic generator and wind energy converter largely complement each other if they are dimensioned correctly, the operating time of the fossil-fuelled generator, and thus the fuel consumption, are reduced (Fig. 7.9).

Some years ago in Germany, an analysis of the cost of power generation by hybrid systems supplying more than thirty mountaineering lodges or hikers' inns in the German and Austrian Alps was carried out. The calculations were also done for radiation conditions in Mexico. The results of this analysis

Fig. 7.10. The Rappeneck (a hikers' restaurant in the Black Forests)

showed that hybrid systems are the most economic solution for supplying these buildings with electric power, not only for the climatic conditions in the German and Austrian Alps, but also under the climatic conditions of Mexico. In both cases, a solar fraction of more than 80% was uneconomical.

Figure 7.10 shows the Rappeneck, a hikers' restaurant in the Black Forest near Freiburg in Germany. The building is situated nearly 1,000 m above sea level, and a connection to the public grid does not exist. The PV generator has a peak power of 3.8 kW and is integrated into the south-oriented slanted roof. The PV system has been operating since 1986, when it was installed by the Fraunhofer Institute without any defect, supplying the building with a great part of its electricity. In addition to the PV generator, a fuel-powered generator and a wind converter with a power of 1 kW can deliver back-up electricity whenever there are unfavorable weather conditions or there is danger of deep discharge of the batteries. Very recently, a fuel cell was installed to complement the energy system.

The Rappeneck is an example of a hybrid system in Germany. But nearly the same concept can be applied for single houses or small groups of houses that have no connection to the public grid in many other regions of the world. In the case of a group of houses, not every house has to have its own PV system or a solar home system. A hybrid system with larger power can supply the whole group of houses like a small central power station.

A new development is the possibility of integrating a fuel cell into the power supply system of a hybrid or other stand-alone PV system. As reported in [95], stand-alone photovoltaic systems need storage to overcome the mismatch between arriving solar radiation and the electric load. To provide continuous power output throughout the year, a back-up power system will also often be necessary. An effective solution to the long-term storage issue of PV systems is based on using fuel cells and hydrogen technology in

Fig. 7.11. The Self-Sufficient Solar House (SSSH) in Freiburg, inaugurated in 1992 (Germany)

connection with PV. A self-sufficient PV-H_2 system consists of a PV array, battery, electrolyzer, hydrogen storage, gas handling unit, fuel cell, and control unit. The battery has the function of short-term storage. The long-term storage is provided by the H_2 gas storage. The fuel cell produces electric energy from the recombination of hydrogen and oxygen to water, which can be recycled by the electrolyzer. In this way, a closed-loop system is realized. Presently, the cost of such a system is still too high for general application, but development is proceeding very rapidly.

After [96], the use of fuel cells in connection with PV systems will open new possibilities for future cost decrease.

From the environmental point of view, the replacement of the conventional part of a typical hybrid system by a fuel cell/electrolyzer will not only lead to a reduction in air and noise pollution, but will also result in the conservation of nature and a better use of renewable resources [96].

This concept was already realized in an experimental house in 1992. After three years of research, planning, and design, the first Self-Sufficient Solar House (SSSH) in the world was completed in Freiburg, Germany [97] (Fig. 7.11). Solar radiation was the only energy source to supply the inhabitants of the SSSH with heat and electricity. The house had no grid connection, and no auxiliary fossil fuels were used. For the heat supply, transparent insulation in the southern building walls was used and a bifacially illuminated flat plate collector provided the domestic hot water.

A PV generator of 4 kWp delivers all the electric energy for appliances and the control system and, indirectly, hydrogen gas for cooking. A lead-acid battery provided short-term energy storage. For long-term storage (summer to winter), an electrolyzer produced hydrogen and oxygen [98]. The gases were stored in two separate tanks. In winter, the gases were either transformed into electricity for electrical appliances by a fuel cell or used directly for cooking and the very small back-up heat demand. The household was equipped with the most efficient (at that time) commercially available electrical appliances for 230 V AC. The project operated for several years in good agreement with simulation results.

7.1.5 Photovoltaic Water Pumping

PV systems can be used for pumping water for irrigation of land, as well as for pumping drinking water (Fig. 7.12). Often these two possibilities cannot be separated clearly. Water which is pumped for land irrigation can be used as drinking water also. Therefore, PV systems for irrigation and drinking water will be described together.

Access to a safe and clean water supply is one of the primary factors in improving the health and quality of life in rural communities. In the developing world, especially in Africa, Asia, and Latin America, a lot of people do not have the option of using clean water for drinking. These remote regions are not connected to a centralized system for supply drinking water. The principal means of water lifting in the developing world are presently the hand pump for smaller demands and the diesel-engine-driven pump for larger quantities. Solar PV pumping can be more appropriate than these technologies in many applications. As communities expand, hand pumping may not be sufficient even if the well capacity is large enough. The disadvantages of diesel generators have already been pointed out. Spare parts and fuel can be difficult or expensive to obtain, and the quality of fuel is often poor due to adulteration, which leads to shorter maintenance periods.

Fig. 7.12. Example of water pumping with PV in Thailand

Fig. 7.13. PV system for water purification (micro-filtration)

PV pumping offers a reliable, low-maintenance water supply, which has zero fuel costs and does not require an attendant to be present during operation. Also, because the time of maximum insolation often coincides with the time of greatest water demand, the supply and demand are well matched. Solar pumping was first introduced into the field in the late 1970s, and since then manufacturers have refined their products to give considerable improvements in performance and reliability. A big advantage of PV water pumping is that no energy storage is needed since the product itself, i.e., the water can be stored.

For example, the importance of PV-powered water pumping systems is described in [99] and [100] for India. In this country, more than 5,000 solar PV water pumping systems from various manufacturers are installed. Among these are nearly 300 pumps for drinking water. Small pumping systems with an array capacity of 300–360 Wp were developed and tried during the 1980s. In 1993, a new program with higher system capacities was introduced. Under this program, users are offered a combination of grants and soft loans. In a recent initiative, the Ministry of Non-conventional Energy Sources (MNES) provided support to the Punjab Energy Development Agency for deploying 500 solar pumps with a 1,800 Wp PV array and a 2 hp DC motor pump set. It is possible to irrigate about 2–3 hectares of land using such solar pumps [100].

Some states in India are planning and implementing larger and larger solar pumping programs, because it is becoming increasingly uneconomical for the state governments to extend the public grid. PV can be useful not only for water pumping, but also for water purification and desalination (Fig. 7.13).

Some of the possible methods are listed here:

- micro filtration,
- ultraviolet irradiation, and
- photovoltaically operated reverse osmosis plants.

In [101], it is reported that according to an analysis made by the World-watch Institute in Washington D.C., U.S.A., 40% of the world population will be living in countries with extreme water scarcity by the year 2025. This expectation makes it necessary for humanity to look for new alternative ways of ensuring a dependable supply of drinking water. The significance of this problem is increasing in the underdeveloped countries as well as in industrialized regions. Desalination of seawater and brackish water is one of these alternatives. In recent years, the process of reverse osmosis has become increasingly important compared with other desalination processes. Some of the reasons for this trend are the low specific energy consumption of this process and the considerable progress made in membrane technology.

7.2 Grid-Connected PV Systems

Grid-connected PV systems always have a connection to the public electricity grid via a suitable inverter (Figs. 7.14 and 7.15), because a PV module delivers only dc power.

Normally there are almost no effects of the PV systems on the grid affecting power quality, load-on lines, and transformers, etc. However, for a larger share of PV in low-voltage grids, as in solar settlements, these aspects need to be taken into account. From a technical point of view, there will be no difficulty in integrating as much PV into low-voltage grids as the peak load of the respective segment [102].

Grid-connected PV systems can be subdivided into two kinds:

- decentralized grid-connected PV systems,
- central grid-connected PV systems.

7.2.1 Decentralized Grid-Connected PV Systems

Decentralized grid-connected PV systems have mostly a small power range and are installed on the roof of buildings (rooftop or flat-roof installation) or integrated into building facades (see Sect. 8.2.1).

Energy storage is not necessary in this case. On sunny days, the solar generator provides power, e.g., for the electrical appliances in the house. Excess

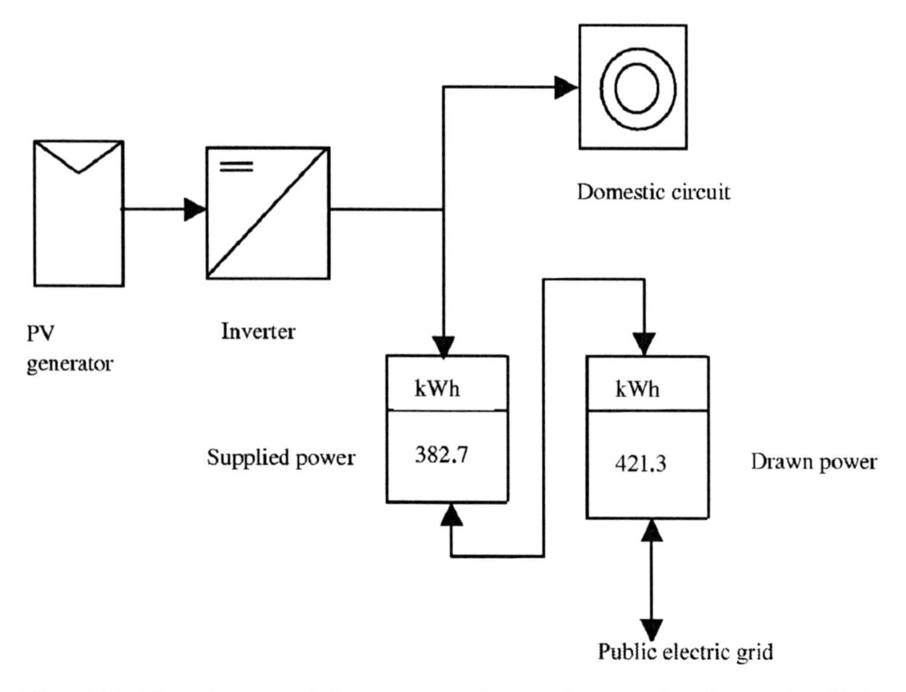

Fig. 7.14. Block diagram of the power supply for a house with a decentralized PV system and grid connection

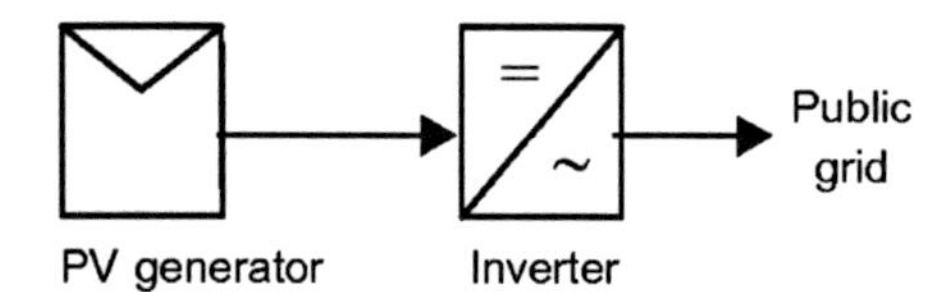

Fig. 7.15. Schematic principle of a grid-connected PV system

energy is supplied to the public grid. During the night and overcast days, the house draws ist power from grid (Fig. 7.14). In this way, the electricity grid can be regarded as a large "storage unit." In the case of a favorable rate-based tariff for PV electricity, as is in force in some countries, it is more advantageous to feed all solar electricity into the grid.

For example in Germany around 80% of the more than 50,000 existing grid-connected PV systems are installed either on the roof of a building or integrated into a building facade. The benefit of the installation of a PV system into or onto a building is that no separate area for the solar generator is needed.

7.2.2 Central Grid-Connected PV Systems

Central grid-connected PV systems have an installed power up to the MW range. With such central photovoltaic power stations it is possible to feed directly into the medium or high voltage grid (Fig. 7.15). Mostly central photovoltaic power stations are set up on otherwise unused land, but in some cases an installation on buildings, mostly on the flat roof of greater buildings, is also possible.

As can be seen in Figs. 7.14 and 7.15 both decentralized and central grid-connected PV systems consist of the following two main components:

- PV module,
- inverter.

The complete system consists of a support structure, cabling, and other conventional components, which will not concern us here. Besides the modules, the most important part of a system is the inverter, which will described next.

7.2.3 Inverter

Inverters are used to convert the DC output of PV or a storage battery to AC electricity, either to be fed into the grid or to supply a stand-alone system. There are many different types of power electronic topologies used in the market. Early PV systems were equipped with thyristor inverters, which are commutated by the grid. Due to their poor voltage and current quality – such inverters have very large harmonics – they have been replaced in the market by self-commutating inverters (with IGBT or MOSFET as semiconductor switches). Only for very large PV inverters in the 100 kW power range thyristor inverters are still used. The first selfcommutating inverters used performed sine-wave modulation of the output current on the primary side and a 50 Hz transformer to adapt the voltage level to grid voltage. Such inverters still contribute a significant market share. Some manufacturers today offer devices with high-frequency transformers instead, thus strongly reducing weight – at the cost of efficiency. In countries where grounding of the PV generator is not mandatory, transformerless inverters are increasingly introduced. They are lighter and more efficient, but particular care has to be taken on EMC[1] and fault current detection.

The serial production of PV inverters was launched in the nineties [103]. Before that time, only a small number of inverters was needed, mostly for stand-alone applications, for example, for residential PV systems. Because there was no connection to the public grid, the standard of power quality in

[1] EMC – electromagnetic compatibility. The inverter is not allowed to send out electromagnetic waves or signals that can influence other electric devices, including the reception of television and radio. On the other side, incoming electromagnetic waves or signals should not affect the correct operation of the inverter.

the system was not as significant as it has to be if the generated power is fed into the grid.

While the first PV inverters were often adoptions of already existing inverters for electrical drives, today's devices are specific developments, taking into account the specific complexity and demands for PV applications. Inverter efficiency is generally more important in PV applications but of particular significance at partial load, as the bulk of the energy is yielded at partial load. Furthermore, with a large-area PV generator coupled to the DC side of the inverter and the public grid on the AC side, stricter standards have to be met with respect to harmonics and EMC.

Some design criteria and functionality of PV inverters are:

- efficiency: well above 90% already at 5% of nominal load,
- cost,
- voltage and current quality: harmonics and EMC,
- overload capability: some 20–30% for grid-connected inverters, up to 200% for short-time overload of island inverters,
- precise and robust MPP tracking (reliably finding the overall MPP in partial shading situations),
- supervision of the grid, safety/ENS[2],
- data acquisition and monitoring.

In order to account for the fact that PV electricity is produced following a typical temporal distribution, in Europe a weighted efficiency measure is used to aggregate the efficiency curve of inverters, the so-called "European efficiency." Good inverters in the kW power range have a European efficiency of 92–96% and up to 98% for large central inverters. A newly developed inverter using the HERIC® (Highly Efficient & Reliable Inverter Concept) topology reaches an efficiency of about 98% [104].

Inverters also perform MPP tracking in order to optimally operate the PV generator. Many different algorithms are used for MPP tracking, trying to cope with three major tasks:

- precision: high precision of MPP tracking needs high-precision measurement components,
- finding the global maximum power output in the case of partial shading, when a local maximum can occur,
- fast enough adaptation of the MPP to changing insolation, e.g., if clouds pass by.

---

[2] ENS (also called MSD) is an islanding prevention device. It has become the de-facto standard for new PV systems. It monitors grid voltage, grid frequency, grid impedance, and ground leakage current, and disconnects the inverter if one parameter is out of bounds. ENS may be either internal to the inverter or an external device installed between the inverter and the connecting point to the grid. Islanding prevention is mandatory for grid connection because if the grid is switched off for maintenance work, all distributed generators have to be disconnected too.

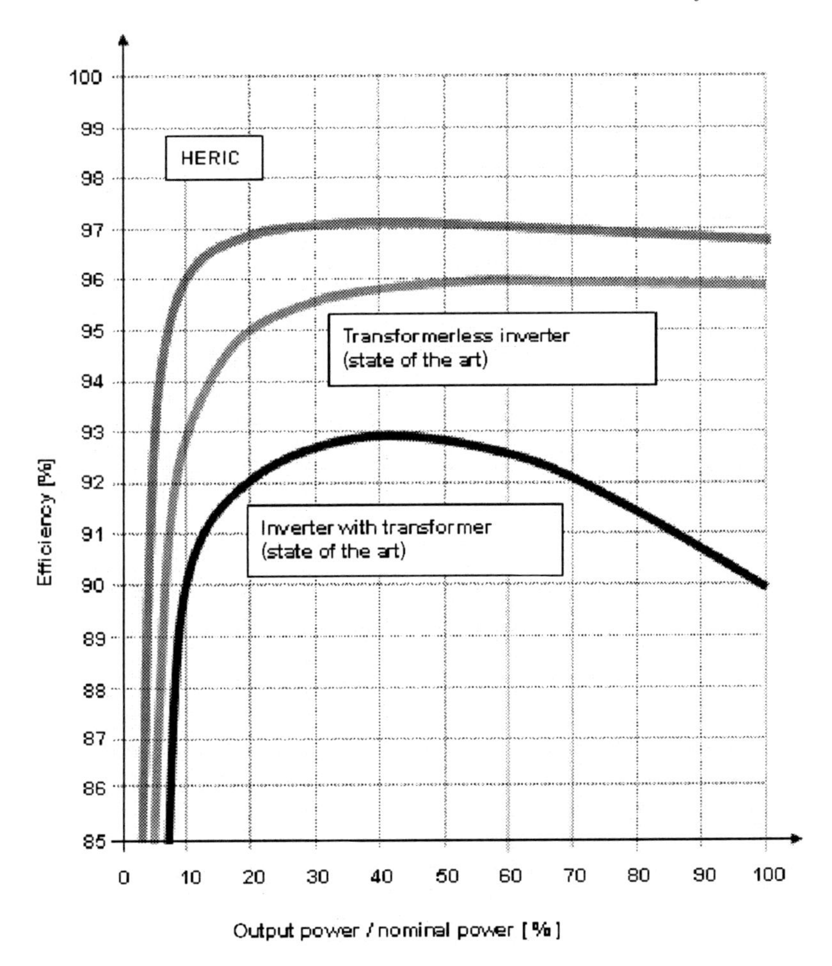

Fig. 7.16. Inverter efficiency curves of three inverters using different topologies

Figure 7.16 shows the efficiency curves of three inverters using different inverter topologies. The top curve is for a new inverter using the HERIC® topology. The second is the efficiency curve of a transformerless inverter and the third the efficiency curve of an inverter with transformer. The last two characterize the state of the art.

8 PV Systems: Installation Possibilities

In this chapter, boundary conditions for the installation and performance of PV modules and generators are described. Starting from general geometrical considerations, we discuss various possibilities of building integration, large power plants in the open field, and tracking and concentrating systems.

8.1 Geometrical Considerations

The optimal tilt angle for large arrays of south-facing modules is approximately the geographic latitude with a tolerance of ± 10 degrees. This arrangement gives the best balance of energy yield over the year. The tilt can also be adjusted for optimal performance in the winter or summer season, by higher or lower tilt. To avoid excessive shadowing, the arrays have to spaced apart by a distance, d, given in Fig. 8.1 in relation to the module width a:

$$d/a = \cos\beta + \sin\beta/\tan\varepsilon\,,$$

where ε can be expressed by the geographical latitude ϕ and the ecliptic angle $\delta = 23.5°$,

$$\varepsilon = 90° - \delta - \phi\,.$$

The criterion for this relation is that the shading angle ε of the preceding module row is equal to the sun's azimuth on solar noon at winter solstice.

The definition of the angles and distances can be seen in Fig. 8.1.

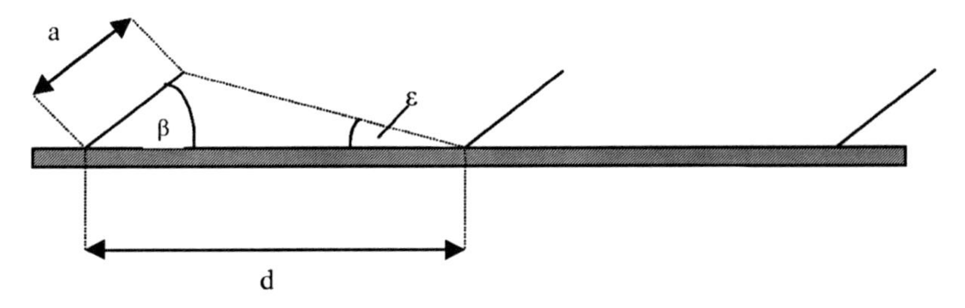

Fig. 8.1. Arrangement of a large number of rows of modules

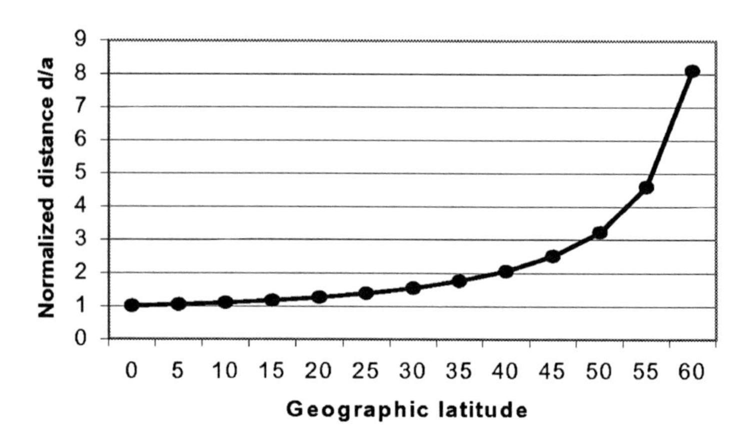

Fig. 8.2. Dependence of normalized distance between rows versus geographic latitude

A modification of this arrangement in connection with plant growth will be discussed in Sect. 8.4.2 of this chapter.

The increase in distance between module arrays with geographic latitude is shown in Fig. 8.2, which is based on Fig. 8.1. It can be seen that the area consumed by a PV generator field increases rapidly with increasing latitude and becomes infinity when the arctic circle is reached (at least when Fig. 8.1 is applied). It should also be considered that the net area required is larger than that, because additional space is needed for access routes, electrical installations, and services.

Much less area is required for an arrangement based on bifacial cells (see Sect. 3.5.2 for the description of a bifacial cell) [105]. Modules consisting of bifacial cells are light-sensitive on both sides. Figure 8.3 shows this arrangement. From non-imaging optics we know that in an arrangement, as shown, of vertical modules enclosed by semicircular mirrors, all radiation falling upon the aperture will be directed onto the modules.

Fig. 8.3. Arrangement with bifacial modules and hemispherical mirrors resulting in more output per module and space saving

The first advantage of this concept is that the available area (e.g., roof) can be much better utilized. For Central Europe, for instance, the yearly yield per total consumed area is 1,053 kWh/m^2 compared to 42,167 kWh/m^2 for conventional modules on racks, because the latter must be spaced apart to prevent shadowing, as shown in Fig. 8.2. A second advantage is that each module receives more radiation than a one-sided module in an optimized arrangement. Under identical conditions, the bifacial module has 1.6 times more output. This includes allowance for reflection losses. In first approximation the collectors are independent of orientation and can be adapted to the shape of a roof. A problem arising with the new setup is that irradiation at the modules is nonuniform in the vertical direction. Since there is no nonuniformity parallel to the horizontal plan, the effect can be counteracted by connecting the cells horizontally in series.

For the economics it is important to note that the expensive module area is replaced by relatively cheap mirrors; additional cost savings result from the smaller area of the generator.

8.2 PV Systems in Connection with Buildings

8.2.1 Advantages and Potential

The installation of PV systems in connection with buildings has important benefits. No additional areas are necessary because the solar generator can be mounted in or on existing parts of a building such as a roof or facade. As described in the following section, the flat roofs are especially suitable for the installation of solar generators. The solar modules can be mounted in optimal orientation and inclination. By installing solar generators on sloped roofs or on facades, only the south-oriented areas of the buildings can be used. In addition, for each of the installation concepts mentioned, the possible shading, e.g., by trees or neighboring buildings must be considered.

In spite of these restrictions, the potential for PV power generation on roof and facades is considerable. In [106], an overview of the potential for building integrated photovoltaics is given. It is the result of a study that was carried out within the framework of IEA-PVPS Task 7. Table 8.1 shows the available areas on roofs and facades offering a good solar yield (80%) as well as the corresponding generation potential in TWh per year. The methodology that was used for the study takes into account the relative amount of irradiation on the surfaces depending on their orientation, inclination, and location as well as the potential performance of the PV system integrated into building. A good solar yield is understood to be 80% of the maximum local annual solar input, defined separately for sloped roofs and facades [106].

The potential given in Table 8.1 can also be expressed as a percentage of national electric energy consumption. Table 8.2 gives for some selected countries the amount of electricity consumption in 2000 and the percentage

Table 8.1. Potential of PV on roofs and facades: Available area with good solar yield (80%) and the corresponding generation potential (TWh/y) for different countries participating in the IEA study [106]

Country	Building stock area (km^2)		Generation (TWh/y)
Australia	Roof	422.25	68.176
	Facade	158.34	15.881
Austria	Roof	139.62	15.197
	Facade	52.36	3.528
Canada	Roof	963.54	118.708
	Facade	361.33	33.054
Denmark	Roof	87.98	8.710
	Facade	32.99	2.155
Finland	Roof	127.31	11.763
	Facade	32.99	3.063
Germany	Roof	1,295.92	128.296
	Facade	485.97	31.745
Italy	Roof	763.53	103.077
	Facade	286.32	23.827
Japan	Roof	966.38	117.416
	Facade	362.39	29.456
The Netherlands	Roof	259.36	25.677
	Facade	97.26	6.210
Spain	Roof	448.82	70.689
	Facade	168.31	15.784
Sweden	Roof	218.77	21.177
	Facade	82.04	5.515
Switzerland	Roof	138.22	15.044
	Facade	51.83	3.367
United Kingdom	Roof	914.67	83.235
	Facade	343.00	22.160
United States	Roof	10,096.26	1,662.349
	Facade	3,786.10	418.312

of this electricity consumption that could be reached by using the potential of PV on roofs and facades. In addition, it can be seen clearly that in half of the selected countries the share on the electric energy consumption that can be met by PV power generation only at existing roofs and facades is nearly 50%. In three other countries, nearly a third of the yearly consumption of electricity could be delivered by PV systems that are installed at roofs and facades. Only in Japan the share is relatively low.

Table 8.2. Amount of electricity consumption in 2000 and percentage of electricity consumption that could potentially be reached with PV on roofs and facades for selected countries

Country	Electricity consumption in 2000 (TWh)	Percentage of PV power generation on roofs and facades
Australia	192.58	43.7
Canada	521.5	29.1
Germany	549.21	29.1
Italy	301.79	42.1
Japan	1,057.33	13.9
Spain	209.55	41.3
United Kingdom	358.28	29.4
United States	3,812.00	54.6

The data in Table 8.2 make clear the importance of roofs and building facades for photovoltaics. By using these areas, a great share of the yearly electricity consumption of mankind can be generated without pollution.

PV systems, which are installed on the roof of a building or integrated into a building facade, are always systems for the decentralized supply of electrical power. In principle, the electrical power produced can be used in the building directly. But if government or utility regulations exist for feed in tariffs (see Sect. 11.2.5) and the feed in tariff is higher than the cost of electrical power drawn from the grid then the entire PV electricity is fed into the grid.

In Sect. 6.2, we described the optimal conditions for the orientation and inclination of modules. When a PV system is installed on the roof or facade of a building, the optimal conditions can often not be met. Then the electrical power produced by the PV system is lower than under optimal operation conditions.

Between 1985 and 1995 grid-connected PV systems were installed in great numbers for the first time on buildings in many countries. At this time, some countries started extensive programs for demonstrating and testing the grid-connected PV technology. One of the aims was to prove that PV systems can be installed and operated without any problems on or in existing or newly erected buildings. Also, within these programs the components of grid-connected PV systems, especially the solar modules and the inverters, could be tested in greater numbers under realistic operating conditions. As a result, these components were greatly improved. The most important introductory programs are listed here.

- The 1,000-roof photovoltaic program in Germany; it was started in 1991 and more than 2,000 grid-connected PV systems up to a power of 5 kWp were installed. The program was concluded in 1995.
- The 200 kWp photovoltaic demonstration and test program in Austria.
- The 70,000-roof photovoltaic program in Japan.

While at first single solutions dominated for nearly each building, now numerous technically proven standard solutions are available. Usually the PV systems installed within the above mentioned programs had a power range of around 5 kWp. This is equivalent to a solar generator area of nearly 50 m^2. Mostly, the PV systems were installed on the roof of single family houses. In Europe, such houses do not have more area with good orientation and inclination. PV systems that are integrated into building facades often have higher power because the usable areas are larger.

The installation or integration possibilities of a solar generator in buildings can be subdivided by building surface or type of installation:

a) By building surface
   - Installation on a flat roof
   - Installation on a sloped roof
   - Installation on a facade
b) By type of installation
   - Mounted on racks (e.g., on flat roofs)
   - Integrated into the roof or facade; in this case, the generator replaces part of the building envelope
   - Mounted at a distance of several cm above the building surface (for better cooling of the modules)
c) Other possibilities of integration such as sun shades

All these alternatives have their own boundary conditions influencing systems technology and performance, which will now be described.

8.2.2 Installation on the Roof

The installation possibilities for a solar generator on the roof of a building depend on the type of roof, i.e., a sloped or flat.

Installation on a Sloped Roof

By using of specific – today mostly standardized – supports, the modules are installed at a distance of around 5 cm from the existing roof tiles. The space between the solar modules and the roof tiles allows cooling of the solar modules (natural convection). The cable connection between the solar generator and the inverter, which is mostly located inside the building, is fed through the roof and has to be protected against the influence of weather (e.g., rain and heat) (Fig. 8.4).

Fig. 8.4. Example of an installation on a slope roof

Fig. 8.5. Example of an installation on a flat roof

Installation on a Flat Roof

The solar modules comprising the solar generator are mounted on a support structure on the surface of the flat roof (Fig. 8.5).

Mounting frames available on the market today have high stability. They also do not violate the integrity of the roof of the structure, and in most cases it is also not necessary to puncture the roof for the cable connection between the solar generator and the inverters because the inverter can be located directly underneath the solar modules. The inverters should be housed in a weather-proof container. The installation of the inverter directly on the flat roof has the additional advantage that the cable connection between solar generator and inverter is very short. This is important because the handling of large dc currents presents many safety problems in connection with arcing.

Another advantage of the installation of a solar generator on a flat roof is that it is possible to optimally mount the solar modules in orientation and inclination.

When solar generators are mounted in rows on flat roofs there has to be a minimum spacing between the rows to prevent shading. In Sect. 8.1 more information about this problem can be found.

8.2.3 Roof-Integrated Systems

There are two main possibilities of integration into a sloped roof (Fig. 8.6):

- Solar modules replace part or all of the normal roofing material. In this case the solar modules have a double function. They perform the function of a roof and, in addition, produce electric energy.
- It is also possible to use special solar roof tiles. These are modules of only a few Wp that are shaped like a regular roof tile. The main advantage of the solar roof tiles is that integration in a conventional roof is possible without problems. The disadvantage of the solar roof tiles is the very high expense for the connection of the great number of solar tiles to one solar generator. Solar roof tiles present the opportunity for very interesting solutions for the integration of solar generators into buildings, even in historic monuments (see Fig. 8.8).

The first roof-integrated solar generators used frameless standard solar modules mounted on a greenhouse structure.

Fig. 8.6. Example of the integration of a PV solar generator in only a part of the roof

Fig. 8.7. Buildings with roof-integrated solar modules at the solar village in Freiburg (Germany)

Later, some companies offered special solar modules that are more suitable for the integration into the roof, and special mounting frames for these modules are used. Figure 8.7 shows, for example, roof-integrated solar modules on the buildings of the solar village in Freiburg (Germany).

The solar roof tiles are not only more expensive than conventional modules, but they also have lower power per area because only part of the tile area is occupied by solar cells. For example, the solar generator at the St. George Church in Burgwalde (Figs. 8.8 and 8.9) consists of 1,028 modules with a power of 5 Wp each. This leads to less output in comparison to a standard solar generator with the same power.

Another fact that reduces the power output of roof-integrated PV systems is that there is no possibility of cooling the modules in the summer by air convection. This applies to modules as well as solar tiles.

The newest development in the field of roof-integrated solar generators are triple amorphous silicon modules, which are produced by Uni-Solar (U.S.A.) (see Sect. 5.3.3). These modules are deposited on plastic films or thin plates of high-grade steel – Thyssen-Solartec® (Fig. 8.10). The two products offer new possibilities for the integration of solar generators into a building. So it is possible to cover the flat roofs of great buildings with roofing material in which a solar generator is integrated. Their output, however, per area is still lower than crystalline silicon because of the lower efficiency of the amorphous cells.

Fig. 8.8. General view of solar roof tiles integrated into the roof of the St. George Church in Burgwalde (Germany)

Fig. 8.9. Detailed view of solar roof tiles integrated into the roof of the St. George Church

Fig. 8.10. Thyssen-Solartec® solar modules

8.2.4 Facade-Integrated Systems

As part of the building, surface facades perform a range of different functions [107].

- Protection from weather (e.g., rain, wind, humidity)
- Sun protection (overheating, glare)
- Use of day lighting
- Protection from noise
- Heat insulation
- Electromagnetic screening

Experience has shown that solar modules are able to fulfill all the above listed functions, and they are also able to meet the properties and the technical characteristics of known building materials. In addition, building facades with integrated solar generators produce clean electrical power.

Facade-integrated solar generators do not have the optimal orientation and inclination. They are practically always vertical. The loss in power output depends on geographic latitude. Assuming a southern orientation of the facade, output increases the further north (or south on the southern hemisphere) the building is located. In Central Europe, a vertical module delivers about 70% of the yearly energy of an optimally oriented module.

Nevertheless, such modern building facades present very interesting examples of solar architecture. The construction of an optically demanding building facade can be combined with an innovative environmentally beneficial power generation. Especially banks, administrations, and environmental organizations have installed facade-integrated PV systems (Figs. 8.11 and 8.12). Depending on the quality of the facade material replaced by PV, such facades can even be economical without regard to the energy production.

Fig. 8.11. The PV facade at the office building of the Fraunhofer ISE, Freiburg

Fig. 8.12. PV facade of the Solar Office at Doxford International, Sunderland, U.K.

The following possibilities of integrating solar modules into a building facade are in most cases offered to the market by some companies:

- standard solar modules can substitute the conventional facade surface,
- custom-made modules with the same dimensions as the surrounding stone or glass facade elements,
- special solar modules consisting of solar cells integrated into insulation glazing,
- semi-transparent solar modules (this means either modules that contain crystalline solar cells with a larger spacing between the solar cells or amor-

phous modules that were made semi-transparent by a special fabrication step),
- solar modules can be equipped with specially heavy glass to act as noise barriers,
- solar modules can also be converted into safety glass,
- solar modules can consist of colored solar cells or have colored spaces between the cells.

The possibilities of integrating modules into a building facade are manifold and some examples will be given here.

An ideal case is the so-called cold facade. The facade containing the modules is mounted at a small distance before the concrete outer wall of the building. Circulation of air is allowed in the space between the facade and the outer wall, which allows the cooling of the modules.

Of course, integration of solar modules into a building facade is also possible in a so-called warm facade. In this case, the modules are mounted directly onto the outer wall of the building. Because there is no space between the outer wall and the facade, no air circulation is possible. Therefore, the temperature of the solar modules can increase and, consequently, the output will decrease.

Very often modules are integrated in only part of the building facade, e.g., as shadowing elements above the windows on the south, or they are integrated at the apron wall. Solar modules can also be integrated into glass roofs or into a glass atrium or into shed roofs, e.g., of fabrication halls. If the modules are semitransparent, they serve additionally to improve daylight distribution inside.

Very elegant facades can be created with so-called structural glazing technology. The glass panes with the integrated modules are fixed with adhesive onto a special support structure. As a result, no frames are visible from the outside. With this technology it is possible to create relatively large uninterrupted glass facades. Structural glazing facade elements can be produced up to a height of eight meters, and they contain all necessary facade elements like windows, apron wall, heat insulation, cabling, and solar modules. In addition, such an element also serves as the inside wall. With the prefabricated elements it is possible to erect the shell of a building very fast.

A special procedure to make crystalline solar cells semitransparent was developed by Sunways (Germany), the so-called power cell. The transparency is created through perforation of the single crystalline silicon solar cell. Solar modules with power cells open up the possibility for interesting architectural solutions, e.g., on building facades. They can also be integrated into glass roofs or windows. The integration of semitransparent solar modules into a glass roof (e.g., into a shed roof) or into the windows offers two practical applications simultaneously – the generation of electricity and the optimization of daylight in the rooms behind the windows.

Fig. 8.13. Integration of power cells into building facades (bakery Härdtner in Neckarsulm, Germany)

Figures 8.13 and 8.14 show two examples of the integration of power cells into the glass facades of buildings.

8.3 PV Sound Barriers

Integration of PV modules into sound barriers along motorways and railways is an interesting alternative to building integration. Photovoltaic noise barriers (PVNB) along motorways and railways today permit one of the most economic applications of grid-connected PV with the additional benefit of large-scale plants (typical installed power: more than 100 kWp). Just as in the case of buildings, no land area is consumed and the supporting structure is already in place. An analysis of existing and planned noise barriers along rails and roads was carried out for the EU [108]. The theoretical and practical potentials were determined. Table 8.3 shows the theoretical potential of the six analyzed countries. It contains the length of all relevant roads and rails and the sum of both. The last three lines of Table 8.3 give the theoretical potential of power capacity of PV systems that could be installed into sound barriers along railways and roads.

Fig. 8.14. Integration of power cells into building facades into the HOKOKU building in Hiroshima, Japan

For the same six countries Table 8.4 shows the share of the theoretical potential, the share of the technical potential, and the share of the short-term potential of PVNB in relation to the respective national electricity production in 1995.

Table 8.4 shows that the share of the theoretical potential of PVNB of the national electricity production in most of the countries listed in the table amounts to around 5 to 6%. Only in Germany is the share 3% lower than the average amount of all six countries.

Table 8.3. Total theoretical potential of installed power for each country [108]

Theoretical potential	CH	DE	NL	UK	IT	FR	All six countries
Relevant roads (km)	1,868	11,013	2,701	10,791	6,830	12,255	45,458
Rails (km)	1,663	6,652	3,065	9,967	4,820	7,850	34,017
Roads [MWp]	2,263	13,183	3,233	12,917	8,176	14,669	54,414
Rails [MWp]	1,422	5,687	2,620	8,522	4,121	6,712	29,084
Rails & roads [MWp]	3,658	18,870	5,854	21,439	12,297	21,381	83,498

Table 8.4. Share (%) of the PV potential of PVNB of the national electricity production in 1995 [109]

	CH	DE	NL	UK	IT	FR	All six countries
Theoretical potential							
Share of PV in electricity generation	5.42	3.00	5.74	4.89	5.80	4.16	4.29
Share of PV in installed power	22.15	16.38	31.82	30.51	18.66	19.87	21.19
Technical potential							
Share of PV in electricity generation	0.04	0.06	0.19	0.01	0.00	0.02	0.04
Share of PV in installed power	0.17	0.34	1.08	0.06	0.02	0.06	1.58
Short-term potential							
Share of PV in electricity generation	0.03	0.02	0.12	0.09	0.08	0.02	0.05
Share of PV in installed power	0.13	0.12	0.66	0.55	0.26	0.09	1.67

The amount of the share of the technical potential that is given in Table 8.4 is cleary much lower than the share of the theoretical potential.

Different noise barrier concepts are applied depending on the shape of the barriers and the degree of noise abatement (Fig. 8.15). All concepts shown in Fig. 8.15 have already been realized – mostly in Germany or Switzerland.

Five of the six examples for PVNB concepts shown in Fig. 8.15 use normal solar cells. The sixth (shown as point four in Fig. 8.15) is another innovative concept using bifacial cells (see Sect. 3.5.2). In a bifacial PVNB, the vertically mounted PV module is used at the same time as a noise reflecting element.

The world's first bifacial PVNB was mounted along a motorway running north-south near the Zurich-Kloten airport in December 1997 [110]. The installation was mounted along a motorway viaduct (Fig. 8.16). The wall elements of a 120-meter-long section of an existing noise barrier were replaced by 50P̃VNB elements featuring integrated bifacial cells. The power range of the PVNB is 10 kWp.

This system demonstrates the highest level of integration possible: The PV module itself actually forms the noise damping structure. Apart from the frame, no additional noise protection materials are necessary. At the location on the viaduct, no requirements for noise absorption exist, thus allowing a reflecting system to be used.

As described in [110], the irradiance of a bifacial installation that runs north-south is simple to calculate. The irradiance on the west side can simply

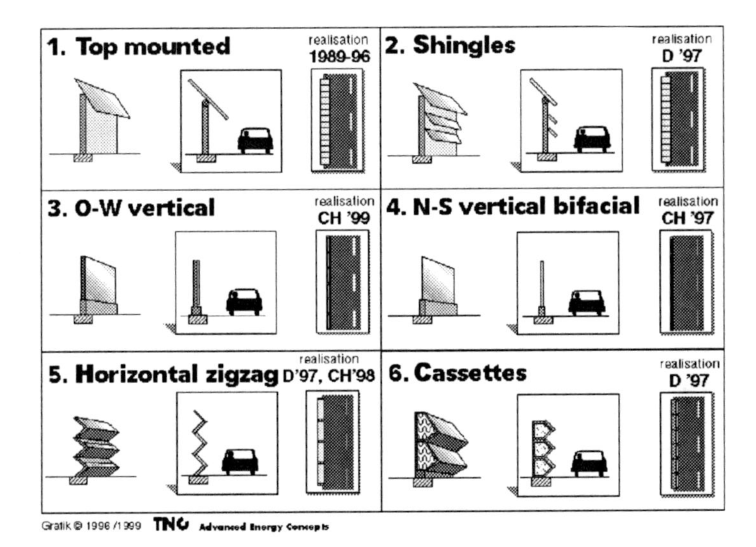

Fig. 8.15. Schematic sketch of different photovoltaic noise barrier (PVNB) structures including integrated PV [108]

Fig. 8.16. The world's first bifacial PVNB near Zurich-Kloten airport

be added to the irradiance on the east side. When this is done, the total irradiance is theoretically 6% higher than an installation oriented to the south with the surface inclined at 45%.

Results of the daily operation of the bifacial PVNB published in [110] show the expected behavior.

Fig. 8.17. Aerial view of the Solarpark Hemau – the solar modules are set up in the open countryside

8.4 Solar Power Plants

8.4.1 Examples of Large PV Power Plants

Large power plants usually have a power of one megawatt or more. They have been installed on large roofs like the Munich exhibition halls, but mainly in the open countryside.

The present growth of the market for PV occurs mainly in the built environment, but with the emergence of rate-based and green tariffs (see Chap. 11) larger PV plants in the open field are also being installed. Especially in Germany some large PV power plants were installed over the last three years.

As an example Fig.8.17 shows an aerial view of the Solarpark Hemau near Regensburg, Bavaria. It consists of forty partial plants, each with a power of 99.36 kWp, so the total power of the PV plant is 3,965 kWp.

8.4.2 PV and Plant Growth

Larger PV plants not integrated into an infrastructure are usually set up in the open countryside. For this purpose, the ground is first cleared of plant growth and then covered with gravel or sand in order to provide a clean and "technical" site [111]. Subsequently, weeds and other plants growing within the PV arrays are exterminated as part of the regular maintenance.

The artificial desert created in manner is tolerable if the PV generator happens to be in a real desert, but when fertile land is used, as is the case in

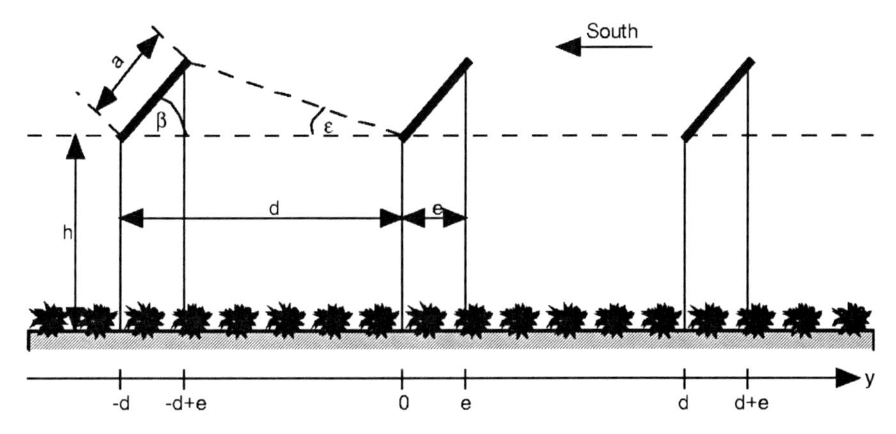

Fig. 8.18. Geometry of the sun farming concept as used in calculations [111]

most of Europe, some disadvantages result. Weeds will grow within the field and must be controlled. This will either require frequent mechanical measures or extensive use of weed killers with associated environmental drawbacks. Even more important is the danger of bad publicity for PV.

This desirable development could easily be endangered by a public perception of PV as a destroyer of landscapes. A corresponding development is currently happening in connection with the deployment of wind turbines. And this discussion has already started about large-scale PV plants. It should also be kept in mind that the full potential of PV can only be realized if the plants can be set up by going beyond the building potential.

We will show here that PV can in fact be compatible with plant cultivation within the same area such that the benefits of both uses of solar energy are additive. Thus, the potential of PV is greatly enhanced. We call this concept sun farming.

The basic principles that PV can be compatible with plant cultivation in the same area were laid out long ago [112]:

The modules of conventional PV plants that are installed in the open field are mounted close to the ground using a similar construction as the installation on flat roofs. Because of this arrangement, the ground below the arrays will be shaded against direct radiation in winter, but in the other seasons most of the light will reach the ground. Thus, the availability of light underneath the collectors coincides with the growing seasons.

If the modules are mounted close to the ground, the penetrating illumination will be highly nonuniform. The solution for making the illumination more uniform is to elevate the modules to a certain level above the ground by utilizing a support structure that causes scant shadowing (Fig. 8.18).

In [111] it has been shown that the radiation level below the arrays is sufficient during the growing season to permit cultivation of most agricultural plants in all parts of Europe and similar latitudes.

In tropical parts of the globe, a variation of the same principle can be applied. Close to the equator the tilt angle is very flat. Then the shade provided by the modules can serve as protection against too much sunshine for sensitive plants underneath. One can imagine that by such means crops can be grown in deserts and, at the same time, solar energy is available for the desalination of water for irrigation.

8.5 Sun-Tracked and Concentrating Systems

8.5.1 Sun-Tracked Systems

As stated earlier, the optimal operating conditions for a PV system are present if the irradiation on the solar generator area is continuously vertical. To realize this in practice the whole solar generator area must be continuously adjusted to the actual position of the sun. Tracking systems can either utilize only one axis or two axes.

- One-axis tracking affects only one angle. The sun is perpendicular to the module only in one plane. The best conditions for one-axis tracking are obtained if the axis is parallel to the earth's axis.
- With two-axes tracking the sun is always perpendicular to the module.

Tracking is mostly connected with concentrating systems, which will be described in the next section, but tracking alone is also sometimes applied. Most tracking systems today are demonstration systems (Figs. 8.19 and 8.20).

Fig. 8.19. Two-axis sun-tracked PV system without concentrating system at the ZSW PV test field Widderstall

Fig. 8.20. PV demonstration building "Heliotrop"

Figure 8.20 shows a demonstration combined-office-and-living building called "Heliotrop."

It is situated in Freiburg (Germany). The PV generator has a power of 5 kWp and is sun-tracked in two axes. That means that the building is following the elevation of the sun during the day. At night it returns to the position at which the sunrise is to happen. In Central Europe, tests have shown that the energy yield of a PV system that is sun-tracked in two axes is around 30% higher than the energy yield of a PV system that is stationary. In locations with more direct sunshine, this portion is higher.

8.5.2 Concentrating Systems

The underlying idea of light concentration is to replace expensive solar cell area with cheaper optical components like mirrors and lenses. Also, as has been explained in Chap. 2, cell efficiency increases with the logarithm of light intensity. Only direct sunlight can be concentrated. Therefore, concentrating systems are best suited for very sunny locations.

At the present time, concentrating PV systems are still at the beginning of commercial utilization. Research and development in the field of concentrating PV systems is going on in several countries. For example, at the First International Conference on Solar Electric Concentrators in 2001 more than 100 attendees from eleven countries took part. Nevertheless, nearly all existing concentrating PV systems are test and demonstration plants. Most of these systems are sun-tracked in two axes. Generally, concentrating PV systems can be classified according to the degree of concentration.

Low Concentration

The concentration factor is 2 to 10 suns. For very low concentration up to two, the system can be stationary (see Fig. 8.21). Concentration up to 10 requires one-axis tracking. Standard solar cells can be used for this concentration range.

The most common PV system with low concentration is the V-trough [113]. These systems mostly use two lateral mirrors adjacent to the PV modules (Fig. 8.21).

In [113] a new system (Archimedes) is described with passive tracking adapted to small decentralized applications. The system is designed for efficient water pumping. Archimedes systems are successfully in operation in Germany, Greece, and Spain. With standard solar cell technology, an annual PV energy harvest of about 2,000 kWh for Central Europe and above 3,000 kWh per installed kWp for southern Europe is realized [114].

Fig. 8.21. Small PV system with low concentration at a schoolbuilding in Freiburg (Germany) without suntracking

Medium Concentration

Medium concentration is 10 to about 60 times. It can still be handled with one-axis tracking. Solar cells used for this range are crystalline silicon cells specially modified to operate at higher currents.

An example for a PV system with medium concentration is the EUCLIDES-THERMIE Demonstration Power Plant situated in Tenerife. The plant was fully deployed by November 1998 and connected to the grid by April 1999 [115]. The plant consists of fourteen parabolic throughs tracking the sun with one N/S-oriented axis parallel to the ground. The collectors are linear parabolic mirrors of $1.8\,m^2$ net aperture made of shaped aluminum plates covered with three types of reflector films. The receivers are 138 concentrator modules made at BP Solar and each containing ten concentrator silicon buried grid solar cells [115] (see Chap. 3). The heat is removed by passive heat sinks, to which the modules are glued. The theoretical peak power of the plant is 480 kW, but the true nominal peak power is 440 kW.

High Concentration

High concentration covers the ranges from 100 to about 1,000 suns and can only be accomplished with two-axis tracking. Because of the high concentration ratio, the cost of the solar cells is a small part of the system cost. Therefore, very expensive high-efficiency cells can be used, mainly special crystalline cells of multilayer tandem III–V cells. These cells reach efficiencies above 30% at high concentration.

PV systems with high concentration mostly use parabolic mirrors or Fresnel lenses to concentrate the sunlight onto the solar cells [116].

One of the biggest high concentration systems is the Amonix/APS 100 Kilowatt high concentration PV installation in Arizona [117]. This installation represents the first large-scale multi-kilowatt deployment of high concentration PV technology in the world. The system consists of four 25-kilowatt sun-tracked PV systems and is based on an innovative integrated structure called MegaModule™ that combines the structural components for aligning the Fresnel lens with the concentrator solar cell. This system uses silicon point-focus solar cells with an efficiency of 26% (see Chap. 3). The efficiency of the system is more than 18% and it has completed large-scale field testing in several geoclimatic zones and is optimized for long-term twenty-year reliability.

Almost all PV test and demonstration systems with high concentration are operating in regions with high direct radiation such as Australia, Saudi Arabia, or the sun belt of the United States.

9 Environmental Impacts by PV Systems

Environment impacts caused by PV systems can be divided into two categories:

- impacts from production of the components of a PV system;
- impacts from the daily operation of the PV systems.

9.1 Environmental Impacts Due to Manufacturing of PV Systems

Silicon, the most common solar cell material, is a product of the chemical industry. Purification involves many process steps that are, however, tightly controlled and subject to strict regulations. The process is operated as much as possible as a closed loop with recycling of reagents (see Chap. 3 for details of the process). Solar cell manufacturing requires diffusion, oxidation, and contacting steps for which different chemicals are employed (Chap. 6). They are either recycled or disposed of in a very controlled manner. Thin-film modules involve different manufacturing processes, which sometimes employ noxious gases. All these steps are tightly controlled and no accidents have ever occurred in solar cell manufacturing.

Cells or modules that are damaged during production are recycled (see Sect. 9.5) into the process.

The inverter is no different from other electronic equipment and is manufactured under the same standards of environmental protection. Therefore, it can be stated that there are no environmental impacts from the production of PV systems.

Energy expenditure and carbon emission by production of PV systems is discussed in connection with energy payback time below.

9.2 Environmental Impacts from Operation of PV Systems

Normally PV systems do not have any effect on the environment from their operation. They do not emit noise, solid waste, or gases that could harm the environment. In this way, PV systems and the electricity they generate make an important contribution to the protection of the environment. This applies to industrial and developing countries. For example, in industrial countries each kWh that is generated by PV plants avoids an output of the greenhouse gas carbon dioxide. The quantitative avoidance of specific greenhouse emissions depends on the actual energy mix in each country. For example, in Germany each kWh generated by PV systems avoids nearly 650 grams of carbon dioxide, but in this case the percentage of brown and hard coal at the primary energy level is relatively high. In some other countries like Norway or Switzerland, a great part of electrical energy is generated by hydro power plants, and so the avoidance of specific greenhouse emissions by each kWh generated by PV systems is clearly lower. Another point of view is that in developing countries PV systems often are the only alternative to the use of fossil fuel, like liquid gas or fuel oil. In this case, the avoidance of specific greenhouse emissions is extremely high.

Since, as mentioned above, PV systems do not generate emissions during operation all their life cycle emissions of carbon dioxide are indirect ones. They result from the manufacturing process and can be calculated by dividing the total amount of emissions generated during the manufacturing of all needed components of the PV system, including possible replacements, by the total energy produced during their lifetime [118].

In case of defects or wear-out of system components during operation, the damaged components have to be replaced with new ones. If the damaged components are not repairable, they should be returned for recycling (see Sect. 9.5). (In developing countries this may be difficult.) It should be mentioned here that the most durable part of a PV system is the solar module itself.

9.3 Energy Payback Time

Energy payback time is defined as the time the energy source – in this case the PV module – has to operate in order to recover the energy consumed for its production. In the strict sense, only renewable energy sources can have an energy payback time, since all other means of energy generation need an input of primary energy. In [119] all past evaluations have been adjusted to be comparable.

- Constant irradiation factors were used.
- Only the modules, no systems components, were included.
- Production technologies studied were close to present production technologies.

The mean energy payback times for the different technologies are listed in Table 9.1.

Table 9.1. Mean energy payback times for PV modules of different technologies [119]

Technology	Mono-Si	Multicrys. Si	a-Si	CIS	CdTe
Mean payback time [years]	7.3	4.6	2.8	1.9	1.5

Table 9.2. Energy payback time (years) for complete PV systems [120]

Technology[1)]	Alsema	Jungbluth	Knapp	Kato
Single crystal	5.2	5.5	7.0	...
Poly crystal	4.4	4.8	...	...
A-Si	3.6	...	...	3.5
CdTe	...	...	...	2.8
CIS	...	...	3.9	...

1) all PV modules with frame

The payback times now have to be compared with the lifetime of the modules. For mono- and multicrystalline cells, most manufacturers give a warranty of 25 years, which can be interpreted as a practical lifetime of at least 30 years. For the newer thin-film technologies, ten year guarantees are customary, but this is due to limited experience. The consequence of Table 9.1 is that PV modules generate between 4.1 and 6.7 times more energy than is required for their production. A further point in this connection is the following: The manufacturing technology today is not optimized for minimum energy input but for lowest cost. Since energy is relatively inexpensive, it can be concluded that the potential payback time is significantly lower than what is shown here.

Table 9.2 contains the results of newer investigations of the energy payback times for complete PV systems [120].

By comparing the data contained in the above two tables, it can to be seen that there is not a great difference even though Table 9.1 shows the energy payback for PV modules only and Table 9.2 shows the energy payback time for complete PV systems. It follows that the energy demand for the manufacture of the solar modules is the main share of the energy demand needed for a complete PV system. The large spread of the numbers also gives an indication of the high uncertainty involved in these studies.

9.4 Land Area Required by PV Systems

The first choice to set up a PV system is on buildings or other man-made structures like sound barriers or shading of parking lots. Recently, some large PV plants were also erected in the open countryside. In this case, it could

Fig. 9.1. Detailed view of the Solarpark Sonnen (Germany)

be argued that they can destroy landscape because they consume a lot of area. Module area alone for 1 kWp is about 10 m^2, but since the modules are inclined on racks with space in between to prevent shading, the actual consumption of space is significantly more, as shown in Fig. 8.2. So far, only land not used for other purposes has been used for PV plants, but, as also described in Sect. 8.2, it is possible to combine agriculture and stock breeding with PV on the same area.

Figure 9.1 shows a detailed view of the Solarpark Sonnen near Passau (Germany), which has been in operation since August 2002. The power of this PV power plant is around 1.7 MWp, and the solar modules are set up in the open countryside. The area of the plant is used as a pasture for sheep at the same time. In conclusion, these examples show that PV does not necessarily lead to the destruction of landscape.

9.5 Recycling of PV Systems

A grid-connected PV system consists of the following components (see also Sect. 7.2):

- solar module,
- inverter, and
- installation material (e.g., mounting racks, cables, etc.).

For stand-alone systems (see Chap. 8), often an inverter is not necessary, but in most cases they need a charge controller and batteries. The electronic and electric components of a PV system such as the inverter, batteries, charge controller, and cable can be recycled by existing and proven technologies without any problems.

Installation materials (for example, mounting supports or frames for facade integration of the solar modules) usually consist of metal (high-grade steel or aluminium), which can be recycled like normal scrap metal.

Special technologies are needed for the recycling of the solar modules regardless of whether the solar modules are made from crystalline silicon, amorphous silicon, or other thin-film material. The increasing use of photovoltaics makes it necessary not only to develop methods for energy and material efficient production of solar modules, but also to create efficient methods for the recycling of solar modules [121]. The recycling of PV modules has been investigated both practically and theoretically. There are no problems with recycling, but only one recycling facility has gone into operation, as described below. Because of the long lifetime of modules, there was simply nothing to recycle.

Already most manufacturers give the guarantee today that all materials that are used for the production of the PV module or systems can be recycled. In the future, it will be necessary to establish a country-wide collection system for defective or old modules. Another possibility is that in the future all manufacturers of solar modules will be obliged to take back solar modules from their own production. For this, the manufacturer or the installation firms, will have to collect the defective solar modules directly from the operators of the PV systems.

9.5.1 Recycling of Crystalline Silicon PV Modules

With a time delay determined by the lifetime of the solar modules, the amount of destroyed or defective modules to be disposed or recycled increases rapidly. If an average lifetime of solar modules of twenty years is assumed, the solar modules that were manufactured in 1985 come up for recycling in 2005 and the modules thath were manufactured in 2003 do not come up for recycling before the year 2023.

As reported in [122], an area of solar modules of about 0.228 km^2 with a mass of nearly 2,300 t was manufactured in 1985. Only ten years later, the manufactured mass of solar modules increased to 8,480 t [122]. This is equivalent to an area of about 0.848 km^2. Most of this are single crystal or poly crystal modules. The percentage of thin-film modules is very low.

At the beginning of the new century, the yearly worldwide production of solar modules increased to about 400 MWp (see Chap. 11). This constitutes a module area of about 4.0 km^2 and a solar module mass of nearly 40,000 t. It must be expected that yearly production will increase at least by about 15% per year.

More than 80% of those solar modules will be made from single or multi crystalline silicon. At present, the manufacturers give a module lifetime guarantee of about twenty-five years, but it can be assumed that the real lifetime

Table 9.3. Classification of crystalline silicon solar modules [123]

Type of module	Embedding material	Front and back side
Standard modules	Film compound (EVA)	Glass – glass Glass – Tedlar™ Glass – Tefzel™ Fluorine polymer – fluorine polymer
	Cast resin (acrylate)	Glass – glass
Facade modules	Film compound (EVA, PVB)	Glass – glass Glass – insulation glass
	Cast resin (acrylate)	Glass – glass
Other module types	As above	As above

of the solar modules is longer than the guaranteed time, maybe about thirty years. Therefore, the time at which recycling of such a great amount of solar modules is necessary will be delayed up to the year 2030 and later.

An additional fact has also to be to considered. In the coming years, modules will be produced with better efficiencies than the modules of today. It is conceivable that fully functioning modules will be replaced with newer and better ones, and the replaced ones will have to be recycled also.

Considering these facts, we come to the conclusion that by 2015/2020 economic and efficient recycling methods for solar modules must be in place.

The presently available crystalline silicon solar modules can be divided into several classes, as shown in Table 9.3.

Depending on the manufacturer, the solar modules contain different materials for the antireflection layer, metallization, cell connections, cable box, sealing materials, and frames. Table 9.4 shows a list of materials that can be found in solar modules and their approximate mass percentages.

The data in Table 9.4 do not contain the frames and their materials. The mass percentage of the frames can amount up to 50%. In Table 9.4, the components and the materials of a so-called standard solar module with an area of $0.5\,\mathrm{m}^2$ is listed. In this case, the mass percentage of the frame is between 10 and 25%. Table 9.4 shows that solar modules contain a multitude of different materials. Therefore, a recycling technology is very demanding.

The manufacturers already have some experience in recycling, because not all modules meet the specifications. Some can be sold at lower prices, but some have to be discarded and are recycled in the factory. Besides the internal recycling, external recycling technologies are presently being developed.

The following methods have been proposed and are at this time the subjects of research and development activities:

The complete separation of the solar modules into their components (e.g., glass, solar cells, metal) with the aim of using the components to manufacture new solar modules. For this purpose, a very detailed sorting of the modules to

Table 9.4. Materials contained in solar modules [123]

Components	Used materials	Approximate mass percentage without frames (%)
Glass (2–10 mm)	SiO_2, Al_2O_3, Fe_2O_3, CaO, MgO, Na_2O, K_2O, SO_3	30–65
Transparent adhesive (1–2 mm)	EVA, acrylate (PVB)	5–10
Solar cell (200–400 µm)	silicon	5–10
Connection material (0.04 × 2–0.2 × 5)	Cu (Sn, Pb, Ag), Al (Mg, Si)	1
Metallization	Ag, SiO_2, Cu, Ni, Al, Ti, Pd, Sn	< 0.1
Antireflection layer	TiOx, SixNy	> 0.1
Doping	B or Al, Ga, In, P or As, Sb	
Cable 1.5–2.5 mm^2	Cu, PVC, rubber, silicone, PTFE	1
Connection box	PVC, PC, PET, ABS, Cu, brass, steel, rubber	0–5
Sealing, gum	silicone, butyl, polysulfide, cyanacrylate	0–10
Back side material	chlorofluorcarbon, polyester	0–10

be recycled is necessary. The sorting must distinguish between the modules of the different manufacturers and, in some cases, even between module types of the same manufacturer.

Conversion of the complete module into so-called ferrosilicon by a high temperature thermal process. Separation into the components is then not necessary. The disadvantage of this recycling process is that the material cannot be used for manufacturing solar cells or solar modules. The resulting ferrosilicon can only be used for the production of steel.

Separation of the complete silicon solar cell from the transparent adhesive and the embedding material by soaking in acids. Laboratory tests carried out by BP have shown that in this way about 75% of the wafers could be reclaimed for manufacturing new solar cells, which, however, had a lower efficiency.

Burning of the embedding material under controlled conditions – in the way the glass panes, metal, and nearly 90% of the crystalline silicon solar cells can be separated from the module. Contamination with metal and plastics does not happen. In a subsequent processing step, the surfaces of the solar cells are cleaned. The resulting solar cells can be used for the manufacturing of new solar modules. Their efficiency is not different from that of newly manufactured cells. The quality of the resulting glass is adequate for the manufacture of modules.

As reported in [122], especially in Germany, tests were carried out to separate solar modules in the existing plants for the recycling of compound glass. But the tests have shown that this is not economical even in the future and that the product did not reach the required pureness necessary for further processing.

Considering the above mentioned information, a recycling of solar modules with the aim of utilizing the resulting components to manufacture new solar modules can be realized by special technologies. Presently, the best chances has a thermal process.

The first worldwide pilot plant using this recycling technology started its operation in June 2003 at Solar World in Freiberg, Germany [124]. The old or destroyed solar modules are placed into a special furnace heated with natural gas. The operating temperature is up to 500°C. At this temperature all plastic materials contained in the modules decompose. Only inorganic materials remain: steel, copper or aluminium, the glass panes, and the solar cells. The metallic materials are introduced into a conventional metal recycling process. The recovered glass is returned to the float glass process. The solar cells – if they are not broken by the thermal treatment – are separated from the metal grid on the surface and the metallic layer on the back side and from the antireflection layer by an etching process. Also, both doping zones are etched off. The broken solar cells can be recycled together with other silicon waste.

The result of this process is a practically new silicon wafer that can be fed into the production of solar cells without problems. The furnace of the pilot plant can process solar modules up to a size of 1.50 to 2.20 meters. For this process, a detailed pre-sorting of the solar modules to be recycled is not necessary. The burning of plastic materials contained in the solar module causes the emission of some harmful gases such as hydrogen fluoride. These gases are decomposed at 800°C and then further purified. In Fig. 9.2, a schematic survey of the process is given.

By using this thermal process, nearly 80% of the primary energy of a normal solar module can be saved. The pilot plant at Freiberg is able to recycle around 150 tons of solar modules per year in a one-shift operation. The optimization of the pilot plant should be finished beginning of 2005. Generally, Solar World is planning a technical solution for the entire photovoltaic industry in Europe for the development of this process.

9.5.2 Recycling of Amorphous Silicon PV Modules

At present and also in the forseeable future, the percentage of amorphous silicon modules to be recycled is and will remain low. In addition, the cell material is a very thin film constituting less than 1% of the module. The transparent conductive ITO (Indium-Tin-Oxide) layer contains a small amount of

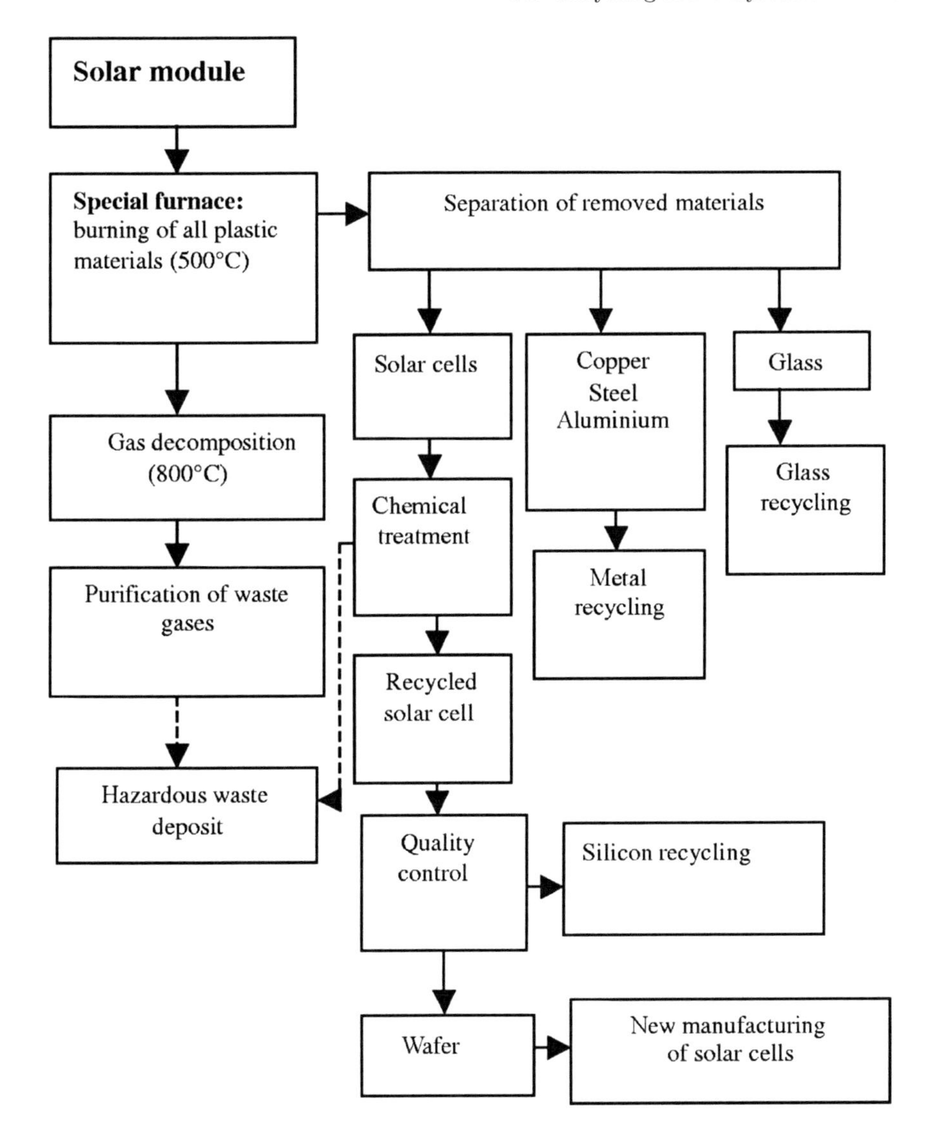

Fig. 9.2. Schematic survey of thermal process for solar module recycling

the heavy metals In and Sn. A good way to recycle amorphous silicon PV modules is to remove the active layers from glass by sandblasting. It is not economical to separate the thin-film materials. Only the glass can be reused. The thermal process described above is also useful for amorphous modules.

9.5.3 Recycling of Compound Semiconductor Thin-Film PV Modules

In contrast to amorphous modules, other thin-film modules contain toxic materials such as cadmium telluride or selenium (see below).

A simple, cost-effective method for recycling CdTe modules has been devised to address the environmental issues confronting the emerging CdTe photovoltaic technology. The method is based on a closed-loop electrochemical approach to rapidly convert defective or spent modules into efficient devices. This method removes and separates the device components and regenerates the semiconductor films on a new panel in a single compact system, leaving no waste materials [125].

9.5.4 Energy Demand for Recycling of PV Modules

Because most of the above mentioned technologies for recycling PV modules are still in the state of research and development work, no reliable information about the energy demand for recycling is available. The thermal process for recycling solar modules described in Sect. 9.5.1 saves nearly 80% of the primary energy demand necessary for the normal manufacturing of a crystalline solar module.

10 Efficiency and Performance of PV Systems

The efficiency of PV systems has two aspects:

- the technical efficiency,
- the economic efficiency.

In addition, the efficiency has to be evaluated separately for stand-alone PV systems and grid-connected PV systems.

10.1 Stand-Alone PV Systems

As outlined in Chap. 7, stand-alone PV systems can have very different purposes and, therefore, they may have different performances and designs.

Technical efficiency can only be compared for similar systems, but generally the following viewpoints are particularly important:

- the reliable supply of the load with electricity during operating time,
- a long lifetime of the stand-alone PV system, and
- expenses for maintenance must be low.

As an example, we describe solar home systems here (see Sect. 7.1.2). For solar home systems in developing countries, both initial cost and service are important. The average monthly cost has been estimated [126]:

- For non-PV-equipped homes, batteries, candles, and kerosene for supplying the home with light and electricity amount to US $6 to 8/month,
- A 50 W PV system with battery and charge controller represents an investment of about US $500. That results in about US $7/month over a time period of about six to ten years (under the assumption of a low interest loan).

At first glance, it is the same amount of money that must be spent per month, but the intangible aspect of comfort is just as important. In terms of the equipment performance, the operational life rather than the efficiency is important, and hence it is the price/service that counts.

For the economics of stand-alone systems, some additional items are needed while others are not necessary compared to grid-connected systems:

- batteries and charge controllers are additional components,
- cables to the public grid and meters are not needed.

The expenditure necessary without PV must be set in relation to the cost of electricity generated by the PV system. Under these conditions, a cost of US $3 to 5 per PV-generated kWh is acceptable.

10.2 Grid-Connected PV Systems

When there is an attractive feedback tariff, the main purpose of a grid-connected PV system is to generate an optimal amount of electric energy over a given time, usually one year. This amount depends on the irradiation at the location in which the PV system is operating and on the quality of the components used. Another possibility is to supply the local load first and to feed into the grid only when there is a surplus. With recent grid failures, a third option of PV becomes of interest, namely, to function as an energy supply for a back-up system. In the following sections, only the first utilization, maximum power production, is considered.

10.2.1 Final Yield

For grid-connected PV systems the energy yield (E_{USE}) measured by a separate AC meter at the inverter exit is basic for all further considerations. Its dimension is kWh. To have the possibility to compare different PV systems at the same operating location, the energy yield is divided by the nominal power (P_N) of the solar generator (kWp), which is given in the data sheet for the solar module. In order to obtain the nominal power of the generator, the number of modules of the generator has to be multiplied by the nominal power of a single module.

The result is the standardized yield or, better, the final yield (Y_F):

$$Y_F = E_{USE}/P_N (kWh/kW_p) , \tag{10.1}$$

where:
E_{USE} = energy yield measured by a separate AC meter (see above),
P_N = nominal power of the solar generator.

10.2.2 Performance Ratio

The performance ratio allows comparison of PV systems at different operation locations.

It is defined by the following relation:

$$PR = 100 \times [E_{USE}/(\eta_{STC} \times E_S)] \% , \tag{10.2}$$

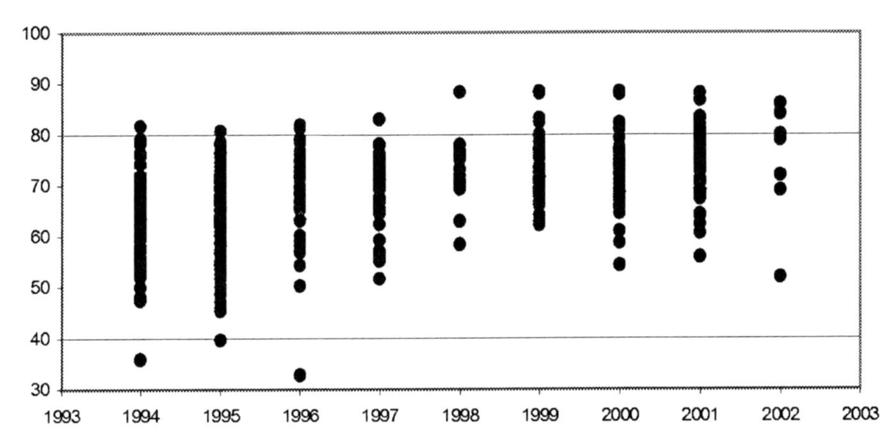

Fig. 10.1. Development of the performance ratio for grid-connected PV systems in Germany

where:
η_{STC} = efficiency of solar modules under Standard Test Conditions (STC) – see "Maximum Power Point (MPP)" in Sect. 6.1.2.
E_S = irradiation at the solar module area during the considered time [kWh/m^2], or quite simplified:

$$PR = Y_F/Y_R\,, \tag{10.3}$$

where:
Y_F = final yield [kWh/kWp],
Y_R = irradiation at solar module area during the considered time [kWh/m^2] divided by the irradiation under STC (1,000 W/m^2).

Y_F and Y_R have the dimension "time" and can be interpreted as full-load operating hours, as nominal irradiation hours. The Performance Ratio (PR) is a dimensionless figure as a measure of system efficiency. In practice, it is specified in percent and describes the effectiveness of the PV system compared with a PV system that operates under nominal operation condition without any losses.

Figure 10.1 shows the development of the PR in Germany during 1993 to 2002. The number of PV plants in each reported year is variable. For example, in 1994 more than fifty PV plants were involved, and in 2002 less than ten. Two trends can be seen. At first, Fig. 10.1 shows a slowly increasing performance ratio from 1994 to 2002. And secondly, there is a wide fluctuation in the maximum PR between the individual PV systems – from less than 60% up to nearly 90%.

The following reasons, among others, can be given for those differences:

- bad quality of the installation of PV systems – for example, the area of the solar generator is temporarily or partially shadowed, especially during times with high irradiation;

Table 10.1. The average performance ratio for grid-connected PV systems in some selected countries [127]

Country	1994	1995	1996	1997	1998	1999	2000	2001	2002
Germany	65.0	64.9	68.0	69.0	73.7	73.6	71.9	73.3	73.4
Italy	58.1	48.2	58.1	55.6	58.6	62.0	67.6	68.6	74.2
Japan*	...	74.0	76.0	70.0	70.0	71.0	71.0	69.0	67.0
Switzerland I**	70.0	70.0	69.0	65.0	63.0	64.0	72.0	...	...
Switzerland II***	69.0	69.0	68.0	71.0	69.0	68.0	68.0	63.0	...
Switzerland III****	...	...	66.0	72.0	70.0	68.0	68.0	66.0	...

* Average annual
** PV systems installed 1989–1992
*** PV systems installed 1992–1995
**** PV systems installed 1995–2000

- the components of the PV system are of bad quality;
- problems with the inverter efficiency; and
- problems with the operation of the PV systems; most often reported are defects and downtime of inverters.

Table 10.1 gives an overview of the development of performance ratio in some countries.

For each of these countries a similar graph as shown in Fig. 10.1 can be drawn. Nearly all of these graphs will show equal trends: a slowly increasing average of performance ratio between 1994 and 2002 and a widely fluctuating performance ratio between the individual PV systems in each of the listed years. The reasons are also quite similar. They are mentioned above.

It is generally accepted that a PV system is of good quality if the performance ratio is higher than 70%. In Table 10.1 it can be seen that PV systems installed after 2000 often reach this value.

Because in Table 10.1 the average values for the performance ratio of greater number of PV systems are given, the performance ratio of some single systems may be higher than the average. So PV systems that have a performance ratio value of nearly 80% are possible.

The reason for the low performance ratio of the PV systems in Italy up to 1998 is on one hand that only a small number of plants were measured and on the other that most of the systems were test and demonstration systems. Their operating conditions changed several times during the year.

If the performance ratio of a grid-connected PV system is known, the final yield can be calculated by using the following formula:

$$\mathrm{Y_F} = \mathrm{Y_R} \times \mathrm{PR} \times \mathrm{SH_F}\,, \qquad (10.4)$$

where:
$\mathrm{Y_R}$ = irradiation at solar module area during the considered time [kWh/m^2],
PR = expected performance ratio according to data sheet,

Table 10.2. Possible final yield at selected locations depending on performance ratio

Location	Irradiation at module area (E_{IR})[1]	Possible final yield with PR 75%	Possible final yield with PR 80%
Berlin[2]	1,143	823	878
Zurich[3]	1,247	898	956
Amsterdam[2]	1,229	885	944
Rome[3]	1,805	1,300	1,386
Madrid[3]	1,910	1,375	1,467
Beirut[3]	1,928	1,388	1,481
Tokyo[3]	1,278	920	982
Sydney[4]	1,810	1,303	1,390
New York[3]	1,619	1,166	1,243
Rio de Janeiro[5]	1,794	1,292	1,378

[1] Average of ten years
[2] Orientation: south; inclination: 40°
[3] Orientation: south; inclination: 30°
[4] Orientation: north; inclination 30°
[5] Orientation: north; inclination 20°

SH_F = factor for the unavoidable shading of the solar generator (e.g., by the horizon or building structures). For for high quality systems in Central Europe a value of 0.96 can be used.

An example of the calculation of the reachable final yield by using formula (10.4) is given in formula (10.5).

If the yearly irradiation at the solar generator area is 1,170 kWh/m^2 and the performance ratio is assumed to be 75%, the possible final yield (Y_F) will be 842.4 kWh/kWp:

$$Y_F = 1,170 \times 0.75 \times 0.96 = 842.4 \,. \tag{10.5}$$

Using formula (10.5) in Table 10.2, the reachable final yields with two different performance ratios are listed for some selected locations.

While in Table 10.2 the optimal possible final yield for some locations are listed, Table 10.3 shows the actual annual final yields for some countries.

At first, it can be seen that the average final yield in the southern countries is higher than in the northern countries, as can be expected from the higher irradiation in those countries. More important is the second piece of information: the observable wide range of final yields for each listed country.

We will use three different demonstration programs to test the grid-connected photovoltaics carried out in Germany as an example to explain this situation in more detail. Figure 10.2 shows the distribution of final yield for these three programs in 2001.

Table 10.3. Annual final energy yields in different countries for comparison [128]

Country	Number of systems analyzed	Range of final yield (kWh/kWp)	Average final yield (kWh/kWp)
The Netherlands	10	400–900	700
Germany[1)]	88	400–1,030	700
Switzerland	51	450–1,400	790
Italy	7	450–1,250	864
Japan	85	490–1,230	990
Israel	7	740–2,010	1,470

[1)] Only systems from the 1,000 Roof Program installed between 1991 and 1995 and located in the northern part of Germany (Lower Saxony)

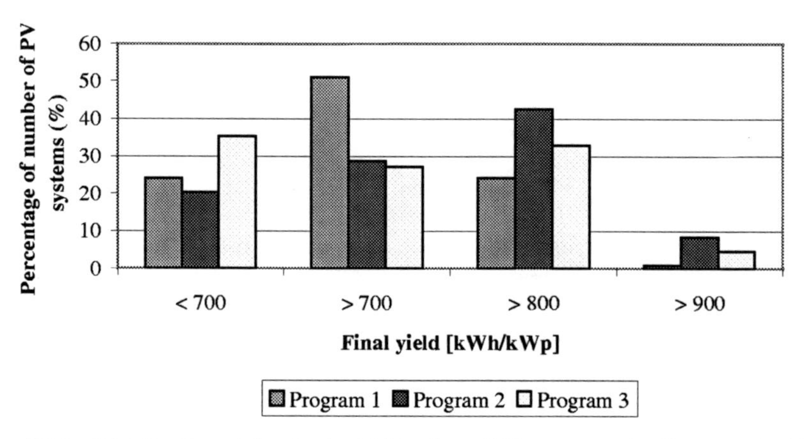

Fig. 10.2. Distribution of the final yield in three PV programs in Germany in 2001

In Fig. 10.2 it can be seen that around 20% to 35% of all PV systems that are involved in the three programs have reached a final yield of less than 700 kWh/kWp. Assuming an average yearly irradiation at the optimally oriented solar generator area (azimut 0° and inclination 30°) of 1,150 kWh/m^2 for the year 2001 in Germany and a performance ratio of 75% (this can be assumed as state-of-the-art), we can calculate the reachable final yield using formula (10.4):

$$\mathrm{Y_F} = \mathrm{Y_R} \times \mathrm{PR} \times \mathrm{SH_F}\,,$$

$$\mathrm{Y_F} = 1,150 \times 0.75 \times 0.96 = 828\,\mathrm{kWh/kWp}\,.$$

The conclusion of this calculation is that all PV systems that reached in 2001 a final yield lower than 700 kWh/kWp do not operate optimally. The reasons for such a reduced final yield belong to the categories mentioned below

- Insufficient quality of the components used. This refers to the solar modules, as well as to the inverter. In the case of solar modules, a discrepancy is often reported between the declaration of the module efficiency on the data sheet and the real module efficiency. Usually the manufacturers of the modules guarantee a difference of ±5% from the nominal power that is given on the data sheet. But in reality the nominal power often is lower than 5%. This is a problem of quality assurance, and most manufacturers have to realize the importance of this fact. It can be assumed that in the future the problem of insufficient quality of modules will be solved. Inverters operate over a long time in part-load operation, and in this operating state often the efficiency of the inverter is not optimal. Newer generation inverters have a good partial load behavior and operate nearly optimally.
- According to the experience in the daily operation of more than 60,000 PV systems in Germany, Austria, and Switzerland, the most important reason for a lower final yield is the shading of the solar generator. The shadow caused, e.g., by trees or houses in the neighborhood of the installed PV generator may fall only at part of the generator area or only for a relatively short time. In each case, a reduction of the final yield occurs. Although the effect of shading on the final yield is known and is often published, this fact is not considered enough in practice.
- Interruption of the operation of the PV system. This disturbance is mainly caused by failing inverters (see Sect. 11.2).

10.2.3 Possibilities of Quality Control and Control of Energy Yield of Grid-Connected PV Systems

Since some countries have adopted feed-in tariffs for grid-connected PV systems, control of the quality and energy yield have become more important than before, especially for PV systems with greater power. In view of this, some companies and engineering consultants offer different methods for the continuous supervision of PV systems. There is no room to describe all the proposed methods, because they are quite different. To assess the efficiency of a PV system only two measured data are necessary: the irradiation in the solar generator area and the energy yield (E_{USE}) measured at the AC level at the inverter exit (see Sect. 10.2.1). To measure the irradiation in the solar generator area usually a special silicon sensor is used. With these two data and the technical description of the individual PV systems it is possible to calculate the final yield (see Sect. 10.2.1) and the performance ratio (see Sect. 10.2.2).

At present, low-priced data loggers are available on the market. The data logger collects the two above mentioned data and also some others like module temperature, ambient temperature, solar generator output (DC), and wind velocity. The data are transmitted to an evaluation computer via telephone connection. Also, transmission by Internet is possible. For a low-power PV system (maximum up to 5 kWp) a monthly reading of the electricity meter

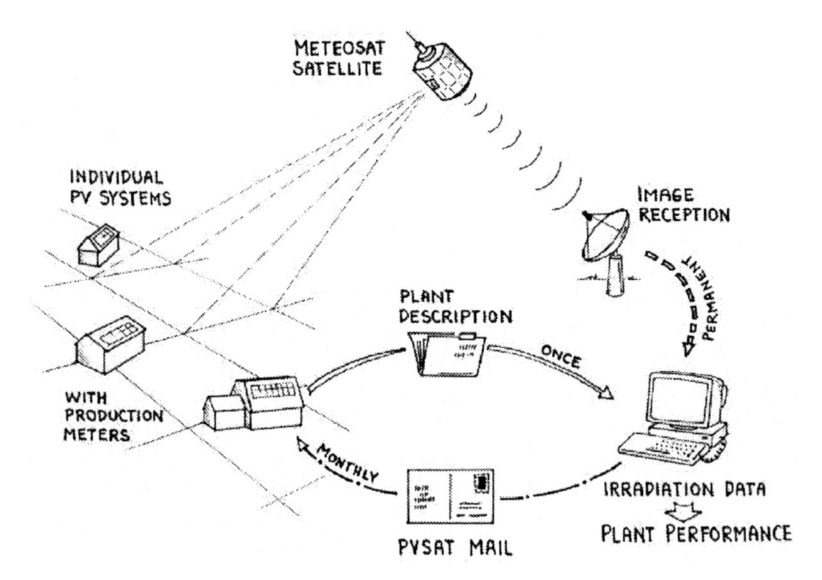

Fig. 10.3. Schematic view of the PVSAT scheme [129]

(DC) of the PV system is enough. Then, the data can be sent for evaluation to a company or an engineering office offering such service, or the owner can evaluate the data himself. For this purpose some companies are offering low-price control units that enable the layman to identify disturbances of the PV system or its components. Especially manufacturers of inverters and data loggers are active in this direction.

Another possibility for evaluation of a PV system is quality control and control of the energy yield by using satellite data [129]. The EU JOULE III project PVSAT has set up a remote performance check for grid-connected PV systems. No additional hardware installation is necessary on site. The site-specific solar irradiation data are derived from satellite images rather than from ground-based measurements (Fig. 10.3). A target yield will be estimated for each individual PV system on a monthly basis. It is reported to the system operators to allow a comparison between targeted and real yield values.

The PVSAT procedure is based on three main components:

- a database of PV systems configuration data,
- a satellite image processor, and
- a generic PV system model.

These data are to be collected once for each participating system.

The continuous reception and processing of METEOSAT images allows for the production of site-specific time series of solar irradiation data for each of the locations. For Mid-European countries, METEOSAT images offer a resolution of $2.5 \times 4.5\,\text{km}^2$. They are available in 30-minute intervals. In consecutive steps, the irradiation on a horizontal plane is converted to the tilted plane irradiation and any local horizon obstruction is taken into account.

At the end of each month, individual yield values are calculated for all PV systems. For this purpose, a generic system model is fed with the configuration data and corresponding irradiation time series. The results of the model calculation are transferred back to the database, from where they are distributed (mailed, faxed, or e-mailed) to the individual system operator. He or she may then compare the estimated production to the real production meter reading.

Also possible for control of PV systems is the combination of data received from METEOSAT with meteorological data collected from a closely meshed grid of meteorological offices. As mentioned above the information about the results of the control can be given by SMS, e-mail, or phone.

10.3 Long-Term Behavior of Grid-Connected PV Systems

Up to the present, only a few studies or analysis dealing with the long-term behavior of grid-connected PV systems had been carried out. Most of them involved only a small number of PV systems or the time period for which the analysis was carried out was relatively short – mostly not more than five to ten years.

Nevertheless, in the following, we will attempt to give a short overview of the long-term behavior of the main components of a PV system.

10.3.1 Solar Module

All known reports and studies about the performance of PV systems came to the coclusion that the solar module is the most reliable component of the PV system. At present, most of the manufacturers give a lifetime guarantee of twenty to twenty-five years for the modules. They guarantee a power of 80% of the nominal power after that time.

The observations of the performance of modules reported here are, of course, not limited to modules in grid-connected systems. Practically all data, however, were obtained from grid-connected systems.

The reasons for the few reported failures of PV systems due to the modules can be traced to failure in the module production process related to insufficient quality assurance by the solar module manufacturer such as delamination or failing solder contacts.

As mentioned in [130], degradation of photovoltaic modules follows a progression that is dependent on multiple factors. Information on module degradation has been collected since the early 1970s. So, for example, module performance losses of 1–2% per year were found in systems tested over a ten year period. Data from a multicrystalline module continuously exposed outdoors in open circuit configuration for eight years show about 0.5% per year performance loss [130].

After [131] a statistical comparison of 191 individual module parameters at normal operating cell temperature (NOCT) shows as a result the decrease in $Pmax$ from 39.88 W to 38.13 W. This represents a 4.39% drop in the average maximum power produced by individual modules during the eleven years they have been in operation.

The database for the reported performance losses is very small. In addition, different measurement techniques and analytic methods were used. Therefore, the obtained results often are not comparable with each other. The observed degradations, which may take up to five years before becoming evident, can be grouped into five categories [130]:

Degradation of Packaging Materials

Module package degradation occurs when the laminate package is damaged or packaging materials degrade during normal service. Packaging degradation can effect, e.g., glass breaking, dielectric breakdown, bypass diode failure, or encapsulant discoloration.

Loss of Adhesion

Delamination is defined as the breakdown of the bonds between material layers that constitute a module laminate. Field experience has shown that front-side delamination at the glass encapsulant and cell encapsulant interfaces is more common than back-side delamination. Front-side delamination causes optical decoupling of materials that transmit sunlight to the cells, resulting in performance degradation. Delamination on either side interrupts efficient heat dissipation and increases the possibility of reverse-bias cell heating. Higher cell operating temperatures cause performance degradation.

Degradation of Cell/Module Interconnects

Interconnect degradation in crystalline silicon modules occurs when the joined cell-to-ribbon or ribbon-to-ribbon area changes in structure or geometry. "Coarsening," a change in joint structure, occurs as a result of segregation of the metals (SnPb) in the soldering alloy. Coarsening causes the formation of larger metal grains, which undergo thermomechanical fatigue, enhancing the possibility of cracking at the grain boundaries and possible joint failure.

Changes in solder-joint geometry caused by thermomechanical fatigue reduce the number of redundant solder joints in a module causing decreased performance. These changes occur due to cracks that develop at high stress concentrations such as voids and thread-like joints. This leads to increased series resistance as current is forced to circulate through diminished solder-joint areas and ultimately fewer solder joints. Interconnect ribbon fatigue has caused degradation in the past, but there have been no recent observations of this problem in field-aged modules. Characteristics directly attributable to interconnect degradation include increased series resistance in the electrical circuit, increased heating in the module, and localized hot spots causing burns at the solder joints, the polymer backsheet, and in the encapsulant.

Degradation Caused by Moisture Intrusion

Moisture permeation through the module backsheet or through edges of module laminates causes corrosion and increases leakage currents. Corrosion attacks cell metallization in crystalline silicon modules and semiconductor layers in thin-film modules, causing loss of electrical performance. Water can also penetrate into leaking connection boxes at the rear of the modules. Corrosion can interrupt the electrical connection between module and inverter.

Degradation of the Semiconductor Device

Degradation of the semiconductor material itself can also contribute to performance loss in field-aged modules. Crystalline silicon modules now have a long track record of performance stability in the field. This stability, in part, is due to the stability of the semiconductor material (crystalline silicon) used to make the cells. Field experience has indicated that the primary causes for performance loss in these modules have been associated with mechanisms external to the cells such as solder bonds, encapsulant browning, delamination and interconnect issues.

Another form of degradation in crystalline cells is a result of chemically assisted diffusion of cell dopant (phosphorous) to the cell surface. High concentrations of phosphorous, along with sodium migrating from soda lime-glass substrates to the cell surface, have always correlated to low adhesional strength at the cell/encapsulant interface. Furthermore, it has been reported that loss of adhesional strength is exacerbated by exposure to high humidity environments.

Degradation of Thin-Film Modules

Amorphous silicon exhibits an initial degradation due to the Staebler-Wronski effect described in Chap. 5. This effect is well known and is taken into account in the efficiency data given by manufacturers. They quote only stabilized efficiency. Amorphous modules very often exhibit a seasonally cyclical behavior. They degrade somewhat during winter time and are annealed by the high temperatures encountered in summer.

CIGS modules appear to be quite stable according to the few data available today.

Degradation Due to Dirt and Dust on the Module Surface

Sometimes it is reported that dirt on the module surface causes a decrease of the energy yield. In this connection the advice is given to clean the module surface from time to time. To this point it can be said that dirt, e.g., dust or leaves on the module front pane, is washed down by the next strong rain if the inclination of the module is more than 5° or 10°. If the inclination is lower than 5°, that means the modules are nearly horizontal, then cleaning of the modules may become necessary.

The need for cleaning is, of course, strongly dependent on the climatic conditions. In a dry desert climate, periodic cleaning of the solar generators is needed.

In summary it can be said that degradation of PV modules is not a very significant effect of the performance of PV systems. Long-term observations confirm the validity of the twenty to twenty-five year lifetime guarantee of module manufacturers for crystalline modules. A problem with all these data is that they pertain to very early versions of modules that are no longer in production today. As a rule, modern modules are even more reliable.

Today, all module types on the market have to pass very stringent qualification tests. In Europe, the Ispra test is commonly accepted.

10.3.2 Inverter

As described in Sect. 7.2.2, the inverter is a piece of electronic equipment like many others. Failures that are customary in such equipment can occur also in the inverter. Some examples are given in the following:

- defective safety fuses,
- breakdown of an electronic component,
- overheating causing faulty operation of the inverter,
- overvoltage effect on the inverter; all modern inverters have an integrated overvoltage protector, but in some cases such as lightning in the direct neighborhood of a PV system can cause an overvoltage to appear at the inverter.

Studies of the long-term behavior of PV systems have shown that the inverter is the most frequent reason for failure of the complete system. More than 60% of all identified interruptions of PV systems are caused by the inverter.

10.3.3 Mounting Racks and Fixing Materials

The only long-term problem caused by mounting racks and fixing materials is normal corrosion. For the selection of the material used for mounting racks and fixing devices, the electrochemical contact voltage has to be taken into account. If is this not done, a rapid corrosion of the less noble metal will occur.

10.3.4 Cables

Normally the cables used for PV systems have the same long-term stability as cables that are used for other purposes, but because the cables in PV system are partly outdoors, the material should be resistant to UV radiation. Additionally, the cable must be resistant to mechanical strain, e.g., shearing of the cable cover.

A special problem has been reported in some cases in Germany. It has been observed that martens have a special liking for the plastic sheets of the cables of PV systems. The martens sometimes interrupt the connection of the generator with the inverter. The consequence is a total breakdown of the PV system. Maybe this problem occurs in other Central European countries as well.

10.4 Electric Safety of Grid-Connected PV Systems

Of great importance for the large-scale introduction of PV systems is the observance of the legal regulation of the electric safety of this systems. If the regulation or standards are not observed by the installer or during the operation of the PV systems, failures or interruptions can result.

The electrical safety of grid-connected PV systems is mostly regulated by national or international standards. An example is the international standard IEC 364 [132], Electrical Installations of Building (IEC = International Electrotechnical Commission), which regulates electrical installations of buildings. This standard is accepted by many countries. The IEC 364 contains, among other things, requirements to protect against dangerous body currents, protection against short circuit and overload, recommendations for the selection and installation of cables and wires, and recommendations for overvoltage protection. There is no room to deal with the IEC 364 in more detail; more information can be found in the standard.

As mentioned already, there is not a single standardized regulation for the electric safety of PV systems that is generally binding. For the installation of grid-connected PV systems, the national and international standards which apply in the concerned country must be observed. In [132], some universally accepted recommendations for materials to be used for the installation of PV systems are given:

PV Modules

In addition to the established standard specifications for PV modules, the following parameters should also be stated in the data sheets:

- protection class,
- maximum permissible system voltage (Voc at STC),
- maximum permissible current in reverse direction.

Cables and Cable Installation

Installation of PV systems presents unique problems, because it is a DC system. This presents a challenge to the installers, who are almost exclusively used to AC systems. The danger of arcing is much greater even at relatively low voltage.

The selection of cable type and cable size depends strongly on the installation method and the expected maximum ambient temperature. In Central Europe, air temperatures close to the roof may reach up to 70°C, therefore, cables should be rated for at least 80°C ambient temperature. In southern countries or at extreme locations, even this rating may not be sufficient. Material used for the insulation of these cables and wires should be rated self-extinguishing; if they are laid openly, they also have to be resistant to UV radiation and weathering. To eliminate the risk of ground faults and short circuits, the installation should be done in a "ground fault and short-circuit proof" way. This requirement usually means laying separate wires for plus and minus (and a possible center tap conductor) either by using single insulated wires in separate cable ducts or doubly insulated cable, type H07RN-F, or a cable type of similar quality. In systems with a higher risk of lightning strike, these two cables should be placed close to each other to reduce coupling of overvoltages from nearby lightning strikes. If a high risk of lightning strikes does exist, it is recommended to place the cables into metallic cable trunks that are grounded at least on either side. All string cables should be marked for string identification and standard colors for "+" and "–".

Terminals and Junction Boxes

Due to their large number, junction boxes of modules are a likely location to develop an arc due to a failing connection. Therefore, junction boxes and terminals inside junction boxes should be nonflammable. The generator junction box has to be included in the protective measure. It should be made from plastic, because this eases the ground fault and short-circuit proof installation. Where no grounded metallic case is present, no ground fault to it can develop. The requirement for ground fault proof and short-circuit proof installation also affects the arrangement of wires in the junction box. Internal wiring must also use doubly insulated cable, plus and minus sections should be spatially separated or physically separated by an insulating barrier. All in- and outgoing wires shall be mechanically secured, e.g., by cable glands. This ensures that even if a terminal loosens and a wire can leave the terminal, it will be technically held in place and unable to contact another metal part.

Switchgear and Fuses

Between the PV generator and inverter a circuit breaker should be installed such that the PV generator output can be disabled if maintenance work on the inverter has to be done. All switchgear on the DC-side like circuit breakers, disconnection terminals, fuses, etc. must be rated for direct current under the full open circuit voltage (at STC) of the PV generator. Failing switches have caused several incidents in the first European PV pilot plants using voltages higher than 200 V.

Fuses should be rated for at least 130% of the respective nominal branch current to allow for periods of increased irradiance. If higher temperatures are to be encountered, e.g., in an attic or in an outdoor junction box, fuses must be derated. An operating temperature of 60°C requires a 20% derating compared to the value at 20°C.

11 PV Markets Support Measures and Costs

11.1 Market Survey

In the last twenty years, the photovoltaic world market has seen a very dynamic development. The yearly average growth rate was 15% over this time, in the last three or four years even 25–30%. The early development of PV was described in Chap. 1. At the end of 1980s the use of grid-connected systems increased more and more, and in the same way the market for stand-alone systems especially for use in developing countries. As a result, PV has spawned a new high tech industry in various countries all over the world. Figure 1.5 shows the evolution of the PV world market from 1983 to 2003. It can be assumed that the recent rapid development will continue into the coming years.

The production of PV modules in different parts of the world has grown drastically in recent years. Until 1998, the U.S.A. led the world in cell and module production, leadership has now shifted to Japan. The figures for 2003 are given in Table 11.1.

Table 11.1. The production of PV modules in different parts of the world

Country	2003 production in MW
U.S.	104
Japan	364
Europe	190
Rest of the world	83

The growth rate depicted in Fig. 1.5 would not have been possible without a whole bundle of support measures, especially for grid-connected PV systems, introduced by governments, utilities, or political groups. They will be described in detail in the next section.

The market shares of different technologies are shown in Fig. 1.2. Crystalline silicon has a market share of more than 90% and amorphous silicon, so far the only relevant thin-film technology, has been losing market share. There are some reasons for this decrease:

- Presently the investment costs, the price of crystalline PV modules, are nearly the same per Wp as the prices of thin-film modules. But since PV

systems with thin-film modules need more area for the same power, the costs for installing and planning PV systems with thin-film modules are higher in most cases.
- The efficiencies of thin-film modules 4.9–6.9% (amorphous silicon), 6.9–7.6% (cadmium telluride), 8.9–9.6% (copper indium diselenide), are significantly lower than the efficiency of modules from single crystal silicon (12–16.1%) or polycrystal silicon (11–14.5%).
- The acceptance of PV modules based on CIS and particularly CdTe is a problem, because cadmium and selenium and to a lesser extent tellurium are toxic materials, but, as described in Sect. 5.2, this is a more nontechnical problem.

The lagging demand for thin-film PV modules and some technical reasons have led to the shutdown of some thin-film manufacturing plants in the recent past. Nevertheless, it is very possible that the thin-film technology will play an important role in the future, because, besides potentially lower cost, it has a number of advantages in comparison with crystalline PV modules:

- the sensitivity to temperature and shading is lower compared to crystalline silicon;
- greater geometrical possibilities for the design of thin-film modules, aesthetic appearance suitable for architecture:
- thin-film modules can be designed transparently.

11.2 Influences on the PV Market

Whereas most stand-alone PV systems operate economically and therefore do not need support for their market development, a market for grid-connected photovoltaic power will not develop under today's price conditions without a strong commitment by governments and individuals to a clean and sustainable environment [133].

Private activities must be complemented and amplified through measures aimed at an active market development taken by governments, electricity utilities, political groups, and private companies who want to participate in a future photovoltaic market.

In Fig. 11.1 we have made the attempt to categorize some possible measures for a PV market introduction. Most of the listed measures have been tried in Germany [133], but in other countries similar measures have been applied. [134] reports about measures that are used for market development in some European countries, North America, and Australia. In the following, details of some "branches" of this tree of measures shown in Fig. 11.1 are discussed together with other possible actions aiming at an active development of the PV market.

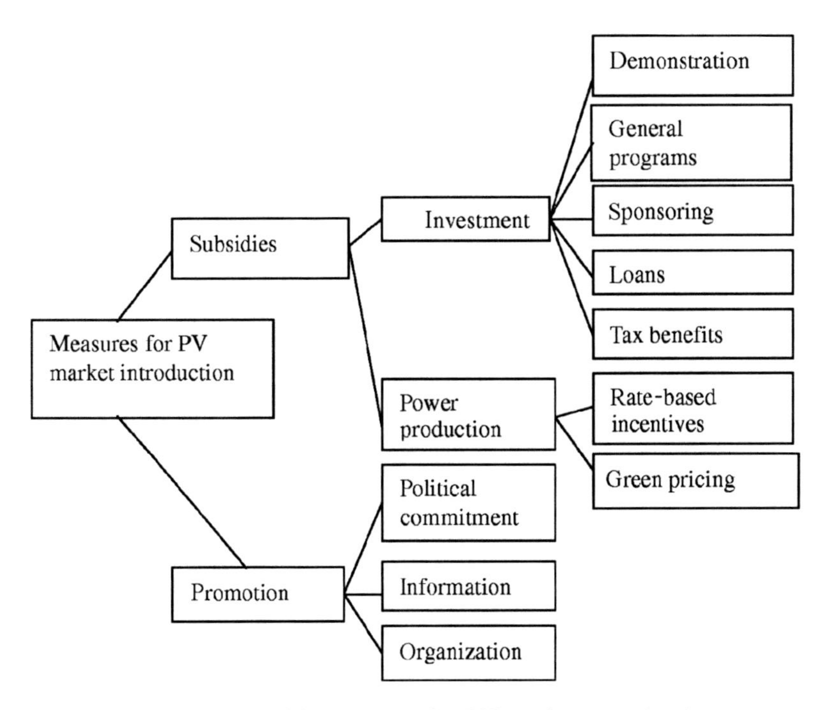

Fig. 11.1. Possible measures for PV market introduction

11.2.1 Demonstration

Demonstration projects show that a new technology is feasible. Often they are pilot projects intended to stimulate the widespread use of that technology. Demonstration projects are mostly subsidized by national governments; in Europe, by the European Union DG XVII or the utilities that operate them.

Some examples will be given in the following:

1. Building integrated systems. As an example, we will describe the so-called Prototype Solarhouse, which was built by NLCC Architects Malaysia and the Fraunhofer Institute for Solar Energy Systems at Freiburg (Germany). Up to now in Malaysia mostly stand-alone PV systems have been installed. Grid connection is exceptional, but photovoltaics is expected to deliver an important contribution to the national power mix in the future. To reach the breakthrough in using photovoltaics, the economic feasibility has to be established and some technical questions have to be solved. The Prototype Solarhouse (Fig. 11.2) demonstrates three different photovoltaic technologies in a typical Malaysian dwelling.

   The house has three staggered roofs. On the highest roof standard crystalline silicon PV modules are installed. In the roof below amorphous silicon thin-film modules are integrated and the lowest roof is partially fitted with so-called solar tiles. The combined power of the three solar

Fig. 11.2. Prototype Solarhouse in Malaysia

generators is nearly 4 kWp. Each one of the described solar generators is connected to the public grid by a separate inverter. In this way, it is possible to monitor the operating conditions of the three different PV systems and to compare results. In addition to the electrical data, climate data were recorded also. With the accumulated data it will be possible to determine an optimal systems concept for upcoming grid-connected PV systems in Malaysia.

2. Other PV systems, for example, the integration of PV modules into noise barriers (see Sect. 8.3).
3. A large number of decentralized photovoltaic power stations were installed to test the efficiency and reliability of the distributed PV systems and their components under real operating conditions for an extended time period by a scientific measurement and analysis program. The best example of such a program was the German 1,000 Roofs Measurements and Analysis Program (German 1,000-Roofs-Program), which was started in 1990. Within this program more than 2,000 grid-connected PV systems (about 5 MWp) were installed on roofs of private houses. The federal and the state governments, which had supported the installation of these systems from 1990 to 1995 aimed to achieve four goals:
   - to harmonize the use of roofs for electricity generation with construction and architectural aspects,
   - to stimulate the users to save electricity and adapt their consumption to the rhythm of solar generation,
   - to optimize all components, and
   - to gain installation know-how.

   Some results of the German 1,000-Roofs-Program are published in [134].

Table 11.2. Monitoring Program of Residential Solar PV Systems in Japan – number of plants, cumulative power, and power per plant of solar generators

FY	1994	1995	1996	1997	1998	1999	2000	2001
Plants	539	1,604	3,590	9,244	15,596	31,4752	57,216	197,216
Power (MWp)	1.9	5.8	13.3	32.8	56.9	114.6	210.4	410.4
KWp per plant	3.5	3.6	3.7	3.6	3.6	3.6	3.7	3.8

A similar program was started in FY[1] 1994 in Japan in order to achieve the target volume of the "basic guidelines." The Ministry of International Trade and Industry (MITI) established the Monitoring Program of Residential Solar PV Systems [135]. This program, operated by the New Energy Foundation (NEF), subsidizes about half the expenditure for purchasing and installing PV systems on the roofs of private houses. Table 11.2 shows the main data of this program.

After [136], in addition to the above mentioned Residential PV System Dissemination Program in Japan, the following demonstration programs were implemented:

- PV Field Test Project for Industrial Use,
- Introduction and Promotion of New Energy at the Regional Level,
- Financial Support Project for Entrepreneurs Introducing New Energy,
- Support Project for Local Efforts to Introduce New Energy,
- Support Program to Arrest Global Warming, and
- Eco-School Promotion Pilot Model Project.

In all these projects or programs the installation of PV systems of different power size is included. As an example there will be given more detailed information about the PV Field Test Project for Industrial Use only. This program started in 1998. The aim is:

- to install trial PV systems using new technology to introduce to the industrial sector, such as industrial facilities,
- to demonstrate availability for the introduction of PV systems by collecting data and analyzing a long-term operation under demonstration tests,
- further standardization and diversified introduction applications toward full-scale deployment of PV systems.

Eligible for subsidy are private companies, local public organizations, and other organizations that are going to install modular-type PV systems and novel applications of PV systems. Half of PV installation cost is subsidized.

PV systems, especially 10–100 kW PV systems, have been installed at schools, welfare facilities, manufacturing plants, warehouses, office buildings, private facilities, and so on.

[1] FY – fiscal year; runs from 1 April to 31 March of the following year.

Table 11.3. Cumulative installed PV power in Japan, total and by four applications (MWp at the end of the year) [136]

Application	1992	1993	1994	1995	1996	1997	1998	1999	2000	2001	2002
Off-grid domestic	0.1	0.2	0.3	0.3	0.3	0.4	0.5	0.5	0.5	0.6	0.6
Off-grid non-domestic	15.3	19.2	23.3	29.4	35.9	44.9	52.3	56.2	63.0	66.2	72.0
Grid-connected distributed	1.2	2.3	5.1	10.8	20.5	43.1	77.7	149.0	263.8	383.1	561.3
Grid-connected centralized	2.4	2.6	2.6	2.9	2.9	2.9	2.9	2.9	2.9	2.9	2.9
Total	19.0	24.3	31.3	43.4	59.6	91.3	133.4	208.6	330.2	452.8	636.8

The above mentioned projects and programs have led to a rapid increase in installed PV power in Japan, especially in the last five years, as shown in Table 11.3.

As already mentioned in Sect.1.2.2, PV demonstration projects also were launched in Austria – 200-kWp-Photovoltaic-Program [138] – and Italy – The Italian Rooftop Program [140].

11.2.2 General Investment Subsidy Programs

This subsidy usually covers a certain percentage of the system cost or a fixed amount per kWp. Investment subsidies were the most common funding model for small PV systems in the first years of the support of grid-connected PV systems.

The advantage of support by investment subsidy is a low administrative expense. A disadvantage is that there is no possibility of controlling the quality of the installed PV power plant. There is information about the amount of power generated by the plant. If, in addition, there is no sell-back rate for PV electricity fed into the grid or the sell-back rate is very low, the plant operator has little motivation to repair defects on the PV systems, especially if the necessary repair is very expensive.

Experience from the 1,000 Roofs PV Program in Germany has shown that the price calculation for a system by the installation companies has a strong tendency toward the maximum of support [137]. The maximum support in the 1,000 Roofs Program amounted to 27,000,– DM per kWp. The analysis of the specific investment cost of the PV plants that were installed in the program has shown that the maximum in the statistical distribution was exactly at 27.000,– DM per kWp.

Nearly the same situation was analyzed for the 200 kWp Photovoltaic Program in Austria. The maximum in the statistical distribution of the specific investment cost was the same as the maximum amount of support [138].

11.2.3 Sponsoring

A special and very interesting measure is sponsoring. From our point of view, the most important motivations for sponsoring are ecological commitment and image. Out of this motivation, utilities, environmental organizations, and the administration in some countries have given grants in different magnitude for the installation of grid-connected PV systems, mostly with small power at public buildings like schools, churches, or other buildings. A PV system is a visible and lasting sign. It can transport the message: "Look what we do. With this PV system we are making a contribution to protect the environment. We are making a contribution to the development of the PV technology, too. Take this as an example to follow and install a PV system at your own home or office building also."

In addition, a well operating grid-connected PV system offers a possibility to inform the general public about photovoltaic technology, especially if the PV system is connected to a visualization or information board showing the actual operating data of the PV system, as well as giving general information about PV technology. In this way, a grid-connected PV system can assist the dissemination of the PV.

In the following, we will give three different examples of sponsorship.

PV at Schools

During recent years in Germany, some utilities have launched programs for the dissemination of knowledge of PV technology. They supported these programs, mostly for schools, with good funding. The utilities were convinced that schools are the ideal place to begin using alternative energy resources as investments in the future.

At present in Germany approximately more than 1,500 school buildings are equipped with grid-connected PV systems. Most of them have a power range from 1 kWp to 2 kWp, but PV systems with a peak power of more than 20 kWp were installed also. With these PV systems the possibility for practice-oriented work in the field of photovoltaics is provided. Having a PV system on the roof of the school building gives teachers the added bonus of using solar energy in their science curriculum. In many cases, teachers and pupils have participated in the installation of the PV systems. They determined the best location for the system at the school building, or they mounted the PV modules on the roof. Most of the PV systems were installed on a flat roof. In addition to installing the PV systems, in a lot of cases the pupils monitor the daily operation. Using computer software, together with their teachers, they can evaluate the efficiency of their PV systems (final yield or performance ratio – see Chap. 10) and compare it with the output at other schools. In this way, the pupils are learning some practical aspects of photovoltaics and getting a lot of knowledge about its theoretical foundation.

A similar project has been launched in Switzerland, where PV systems were installed on technical colleges [139], and also in Italy [140].

In [141], a program is described that provides hands-on experience. It gives problem-based learning opportunities to all Hong Kong primary and secondary schoolchildren. Intended to raise their awareness and real understanding of the contribution of renewable energy technologies to everyday life, the program has a further objective to expand the experience of local construction professionals in renewable energy technologies.

In another example, the Sacramento Municipal Utility District (SMUD) has also installed grid-connected PV systems on school buildings [142].

In [143], the largest school-based solar energy network in the U.S. is described. Eight Chicago public schools have installed 10 kWp utility-interactive photovoltaic systems. The school-based solar electric installations help to educate and raise awareness among schoolchildren about protecting the environment by using renewable energy. And at last, the Solar Program in the U.K. should be mentioned. In this program, nearly 100 PV systems were installed in schools and colleges throughout the country [144].

PV at Churches

The well-known project to install grid-connected PV systems at churches was launched from the Deutsche Bundesstiftung Umwelt (DBU). The program is called Parishes for Solar Energy. It uses the important role of church parishes to set an example for the general public.

In the frame of this program financial backing is given for solar thermal and photovoltaic demonstration plants on church properties. In contrast to some other programs supporting solar energy, the given grant includes the technical plant (solar thermal or PV system), as well as a visualization board and the public relations activities that are necessary to reach a wider audience.

Fig. 11.3. Grid-connected PV system at the St. Nikolai Church in Leipzig

In the frame of this program more than 600 solar energy demonstration plants, among them more than 500 grid-connected PV systems, were installed. The average power of these systems is about 4 kWp [146]. Altogether in Germany at churches or church properties a PV power of more than 2 MWp is installed today.

Figure 11.3 shows an example of a PV system from this program. The grid-connected PV system is integrated into the roof of the church St. Nikolai, one of the oldest churches in the city of Leipzig. The PV system has a power of nearly 5 kWp.

PV at Public Buildings

Public buildings like the head offices of banks, environmental organizations, and utilities are often very suitable for the installation of grid-connected PV systems. Very often these buildings have large glass facades. Instead of the glass elements, special PV modules can be integrated into the south-oriented part of the facade (for more information, see Sect. 8.2.4).

Such a facade-integrated PV system can give a widely visible symbolic sign to the public, can improve the image of the utility or bank, and, as mentioned before, provides a good possibility for the dissemination of information about PV.

11.2.4 Low Interest Loans

A well-known program of supporting the PV by loans is the 100,000 Roofs Solar Program, which was started by the German government in 1999 [137]. Within this program, the government supported the installation of PV systems of a size of 1 kWp and by offering low interest rate loans. The initial conditions at the start of the program included a ten-year, interest-free loan to be repaid in nine annual installments, no repayment being required in the final year. The final 10% installment was waived. The realization of the program was organized by the Kreditanstalt für Wiederaufbau (KfW), a German bank which is 100% owned by the German government.

Participation in the program could be combined with additional local support measures, including combination with some municipal full cost-recovery pricing schemes up to a total assistance of 100%. Due to some confusion concerning details of the technical organization of the money transfers between the banks, the combination of state and federal subsidies, limits of sizes and others, there was a delay at the beginning of 1999, but at the end of this year nearly 4,000 systems with a total capacity of 10 MWp were installed under the program.

Early in 2000, with a new German Renewable Energies Law (see below), feedback rates for PV electricity were increased significantly. So, in addition to the zero-interest program, plant owners were paid US $.50/kWh for

Table 11.4. 100,000 Roofs Solar Program: original versus revised goals

	1999	2000	2001	2002	2003	2004	Total
Goal – original plan (MWp)	18	27	36	51	72	96	300
Goal – revised plan (MWp)	10	50	65	80	95	–	300

PV-generated electricity. Making PV fully economical led to a boost in applications for the 100,000 Roofs Solar Program – during the first four months more than 70 MWp had been applied for. Since this quantity exceeded the original planning significantly, the government placed a moratorium on applications and revised the scheme for the loans quickly. Major changes have been the dropping of the one year free (the tenth year), a lower limitation of the loan maximum for larger systems, and the introduction of an interest rate.

The government also changed the goals for annual installations. Table 11.4 shows the goals of the original program and the goals of the accelerated new program.

The actual conditions (spring 2003) of the 100,000 Roofs Program are the following:

- Private persons, freelancers as well as small- and medium-sized enterprises, could apply for participation in the 100,000 Roofs Solar Program. If the application for participation was approved by the KfW, the participant in the program received a loan of up to 6,230 € per kWp for PV systems with a power of less than 5 kWp,
- For PV systems with higher power, the maximum loan was 3,115 € per kWp.

For the loan there was fixed an interest rate 4.5% less than the usual market rate. The limit of the loan volume was 500,000 euros. The duration of the loans was ten years and the first two years were interest free.

The development of the installation of PV systems that were planned (and realized) with the 100,000 Roofs Solar Program, is shown in Fig. 11.4.

The interest in the 100,000 Roofs Program was very high. Therefore, when the budget of the first program was exhausted, the KfW launched an additional program to support the installation of grid-connected PV systems in Germany. As in the 100,000 Roofs Program participants can get a loan to finance the installation of a PV system. But the interest rate in this program is fixed at 3.95%. The participants in the above mentioned special program also can combine it with the feed-in tariffs of the German Renewable Energies Law.

The German 100,000 Roofs Solar Program ended at 31 December 2003. Because the German government wants to support the installation of PV systems in the future, the feed-in tariffs for PV systems were changed at the beginning of 2004. For more information, see Sect. 11.2.5.

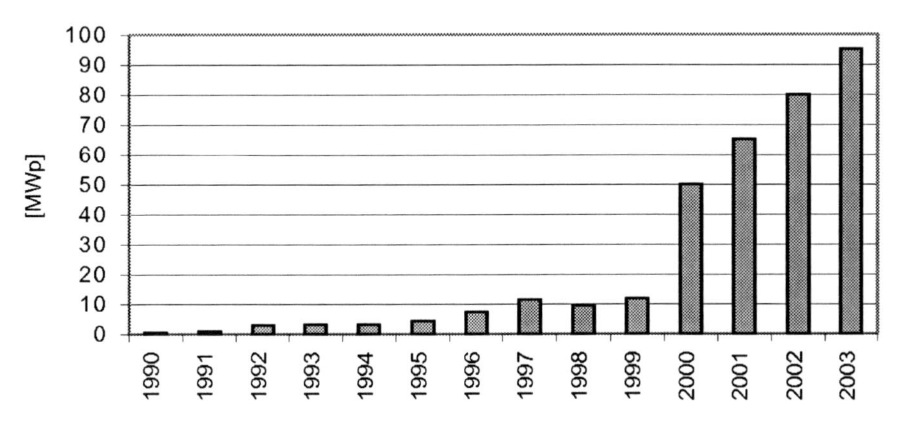

Fig. 11.4. Yearly installed grid-connected capacity in Germany. The influence of the 100,000 Roofs Program starting in 2000 can clearly be seen

11.2.5 Tax Benefits

Some governments, for example, Germany and few states in the U.S. give a tax benefit of different magnitudes for a PV system installed in conjunction with the purchase or construction of a private home [134]. Tax benefits are given in some cases in conjunction with the reconstruction of older private homes, too.

11.2.6 Rate-Based Incentives or Feed-In Tariffs

In contrast to general investment subsidy programs (see above), sell-back rates do not fund the installation of a PV system but its product, the generated kWh by a PV system [137]. Some countries have introduced this way of funding grid-connected PV with great success.

Germany: Since the beginning of 1991, German utilities have had to buy PV electricity fed into the grid at 90% of the average electricity rate of the year before. The mandatory PV electricity sell-back rate in recent years was thus about 0.09 euros. In addition, many local programs with increased sell-back rates exist. This rate was effective for some renewable energies like wind, but much too low to make PV attractive. The Solarenergie-Förderverein (SFV), Aachen, developed a cost-covering rate model for a model PV system of 5 kWp, the total cost, including depreciation over twenty years, is determined each year. In 1997, this was about 1 €/kWh. The utility should pay this rate to the operator of the PV system. The cost of this support for PV is added to the general electricity rate, the increase of which is limited to a given program ceiling, usually a 1% surcharge. This regulation is very important, because this way PV is not subsidized by public funds, but by a levy on electricity prices. Only thus can the rate be guaranteed for twenty years, because it does not have to be approved by Parliament each year. One can also argue that it is a fair regulation, because the cost is allocated to the source of the problem, the CO_2 emissions by electricity generation.

Each year, the rate for a new model system is determined in order to promote competition and price decrease. The owner of the PV system is guaranteed the rate for the year of installation usually for ten to twenty years.

On 22 November 1999, the cabinet of the German government agreed on a significant increase of sell-back rates for PV electricity, leading to a cost-oriented funding for photovoltaics, which was passed as Renewable Energy Law by the German Parliament on 25 February 2000. It replaces the earlier Electricity Feed-in Law and guarantees prices for generation from a wider range of renewable energy technologies than before. The compensation to be paid for electricity generated from solar radiation energy was then at least 99 pfennigs (about US \$.50 or .50 euros) per kilowatt-hour. For the following years, a declining-scale remuneration was established. The reimbursement drops by 5% each year for newly installed systems. This drop is an important part of the law, because it demonstrates that PV is an energy source with a positive cost outlook. The 5% results from the established learning curve for PV. The rate fixed in the year of installation is guaranteed for twenry years. The law stipulates that each owner of a PV system has the right to feed energy into the grid at the legal rate and the utility has to comply. In this way, obstruction by utilities is prevented.

Because the German 100,000 Roofs Solar Program ended at 31 December 2003 (see Sect. 11.2.3), the German government changed the conditions of the feed-in tariffs for renewable energies fed into the grid. The actual conditions for PV follow.

- For **PV systems set up in the open countryside**, the feed-in tariff is 0.457 €/kWh.
- For **PV systems installed on buildings or on noise barriers** up to a power of 30 kWp, the feed-in tariff is 0.574 €/kWh; for PV systems with a power between 30 and 100 kWp, the feed-in tariff for the part of the PV system above 30 kWp is 0.546 €/kWh and for PV systems with a power of more than 100 kWp, the feed-in tariff is 0.54 €/kWh.
- **Facade-integrated PV systems** get an additional bonus of 0.05 €/kWh.

Spain: With Royal Decree 2818/1998 the sell-back rate for PV electricity is fixed at 0.4 €/kWh for systems with a power of less than 5 kWp and 0.2 €/kWh above. The Spanish law has several shortcomings that have so far prevented a strong development of the Spanish market.

Japan: In Japan, the sell-back rate is 24 Yen (0.21 €) per kilowatt hour. This rate is identical to the electricity rate paid by the customer. This arrangement is also called *net metering*. It has been introduced in some U.S. states and some other countries. It is, of course, more effective in regions with high electricity cost like Japan.

Austria: Since 1 January 2003, the feed-in tariff for PV current has been countrywide 0.60 € per kilowatt hour for PV systems up to a power of 20 kWp and 0.47 €/kWh for PV systems above. The feed-in tariff was granted temporarily for thirteen years, and there was a limit of the total installed PV power capacity of 15 MWp. This limit is very low and was exhausted rapidly.

France: France's first feed-in tariff is currently in place after an agreement could be found for the regulation of interconnection and metering with the utility Electricite de France (EDF). For systems installed in the first year, the PV tariff, guaranteed for 20 years, will be 0.15 €/kWh in France and twice that for French overseas Departments and Corsica (0.30 €) and will decrease by 5 percent annually for systems installed in succeeding years. The scheme will cover residential systems up to 5 kWp, non-building systems (such as noise barriers) up to 150 kWp, and commercial and public buildings up to 1 MWp.

The low amount of the feed-in tariff is being combined with a 4.60 € per watt PV rebate offered by ADEME, France's environment and energy management agency, but the incentive is limited in the mainland to a meager total 1 MWp for each of the three years of the program. Connection costs for PV owners, who will need to rent a second meter, will be low.

Portugal: Since 1988, the so-called E4-Program has existed. According to this program, the feed-in tariff is 0.50 € for plants with a power up to 5 kWp. For plants with higher power, the feed-in tariff is 0.45 € per kWh. The feed-in tariff is guaranteed for twelve years and there is a limit in the total installed PV power capacity of 50 MWp.

In Greece, Belgium, and Italy, the feed-in tariffs are comparatively low – around 0.15 € per kWh, and in *The Netherlands* and *Denmark*, the feed-in tariffs are at the same level as the customer tariffs (net metering). In addition to the mostly low feed-in tariffs, the governments of Belgium and Denmark support the investment cost in different magnitudes.

As is detailed in [134], Portugal, Greece, and Italy have had little success with their feed-in tariffs. They failed to provide additional support with planning permits, cheap loans or grid access, or long-term security, as in Spain.

11.2.7 Green Pricing

Some utilities offer a green tariff on a voluntary basis. They offer energy produced by renewable sources, including PV, at a surcharge. The utility installs the systems and in some cases adds its own money to the funds from the surcharge.

11.2.8 Foundation

Customers of utilities pay a contribution to a foundation used for financing PV plants. The utilities typically add the same amount. Foundations are similar to green pricing. Instead of coupling the individual contribution to the electricity consumption, the customers pay a fixed sum.

11.2.9 Solar Power Stock Exchange

The utility trades privately generated PV electricity. The utility calls for bids of PV electricity and selects the most attractive offer. The municipal utility EWZ Zürich introduced this successful model developed by TNC [146]. Another example is BEWAG Berlin (Germany).

11.2.10 Cooperatives

An organization installs and operates a PV plant and sells shares. The shareholders get the returns from the electricity sales. Especially in Germany, jointly owned PV systems often are installed and operated by private cooperatives. In this way, private persons have the opportunity to participate in the ownership of a PV power plant even if they do not have the option of installing their own. In addition, they get the cost benefit from a larger plant.

11.2.11 Green "Utility"

Companies or organizations specialize in producing and selling green electricity. The utility itself may be financed by private shareholders, thus being a jointly owned PV utility, instead of operating a jointly owned PV plant. Customers of the company or organization may be end users or distributors like municipal utilities. In the first case, the company may be called a green utility. In the second case, it supplies the suppliers.

11.2.12 Tendering

Whereas feed-in tariffs (Sect. 11.2.5) pay the same rate to all plants of the same technology, tendering requires a bidding process. Because of that, the administrative burden for the project developer or potential operator of the PV plant increases. After [134], tendering for government monies for green power projects was carried out in France, U.K., and Ireland. Tendering has proven to be far less efficient than feed-in tariffs. An analysis of the results and workings of the U.K.'s tendering mechanism (called the Non-Fossil Fuel Obligation – NFFO) by the author of [134] has shown that only some of the projects accepted during the tendering process were ever realized. This can be attributed to the fact that the tendering process encouraged speculative bidding, with developers submitting bids before having obtained planning permission and at prices that made unrealistic assumptions about the final cost of the projects.

As a consequence, the U.K. is using the NFFO no more. Instead, a mechanism that is like the Renewable Portfolio Standard being used in the U.S. (see next chapter) is used.

11.2.13 Renewable Obligation Order or Renewable Portfolio Standard

In [134], it is reported that thirteen U.S. states, Australia, Belgium, and the U.K. have legislated obligations (called Renewables Obligation Order or Renewable Portfolio Standard (RPS) in the U.S.) that the share of renewable power be increased to a given percentage year by year. The targets vary extensively. For example, in Australia 2% and in the U.K. 10% of the national power mix by 2010 are planned. The targets include only a small amount of power generated by PV.

11.2.14 Installation on Leased Roof Areas

In Germany, a special measure for PV market introduction has been developed. A utility or company, including cooperatives, operates PV systems that are planned and installed on the roofs mostly of private and public buildings. These areas are leased from the owner of the building for at least twenty years. The electric energy generated with these PV systems is fed into the grid and then sold to other utilities that offer "clean" electric power, or to the final consumer.

Especially in the southern part of Germany, but also in Austria and Switzerland, a large number of PV systems with a total power of nearly 4 MWp are installed this way [147].

11.2.15 Political Commitment

The political influence of private groups or political parties plays an important role in improving the boundary conditions for PV and for renewable energies generally.

According to [148], the access to the grid is the first imperative for renewable energies to gain a foothold. Three main types of regulatory policies have been used to open the grid to renewable energies. One guarantees price, another ensures market share (mandated targets), and the third guarantees utility purchase of excess electricity from small-scale, distributed systems.

As an example of the effect of this possible influence on the change of real politics, we refer to Germany. In 1998, after the election victory of "red-green," there was a change in government and a change in the German energy politics also. The new government has launched the 100,000 Roofs Solar Program as well as the new Renewable Energy Law. Both have turned out to be a very big push for the photovoltaics in Germany. Consequently, the production capacity for PV modules and PV inverters, as well as the import of PV modules for the German market increased significantly. This also led to an increase in research and development in the field of photovoltaics. Today, Germany is one of the leading countries in the world in production and use of PV systems and components.

11.2.16 Information

The best technology is useless unless you know about it. General information is disseminated by the public relations departments of public authorities, utilities, companies, and research institutes, as well as by ecological or solar energy associations – many of which publish via the Internet. Market surveys, guides, yellow pages, test and technical journals, exhibitions, fairs, and conferences provide more specific information. Municipalities and utilities give practical advice for PV installations in the framework of their energy consultancy. Last but not least, PV is an educational issue – from general schools to professional training.

Information can have two sides. One side is to provide factual information on the operation, potential, and limits of PV. The other side attempts to give false or negative information about PV. The aim of the misinformation is to make a potential buyer of a PV system feel insecure in his decision. Mainly false information about the cost (too expensive) and the operating conditions of PV systems (only usable for developing countries, unsuitable for industrial countries) will be given. Another argument is that PV is a destroyer of landscape (see Sect. 8.2.2). Especially the daily and annual fluctuation in the solar radiation is used as a knockout criterion for photovoltaics. Last but not least, a very popular argument against photovoltaics is that the energy demand for the production of a PV system is higher than the energy output that can be achieved during the complete lifetime of the PV system (see Sect. 9.3).

It is very important to correct these misconceptions and to provide well founded information about photovoltaics. We hope that this book can contribute to this aim (see Chap. 14).

11.2.17 Evaluation of Market Support Measures

Experience with all the different supporting mechanisms described above allows a conclusion about the best instrument for supporting PV. This is clearly the cost-covering rate model that was developed by the Solarenergie-Förderverein (SFV), Aachen. It leads to the installation of a large number of PV systems with high quality, because it is in the interest of the PV plant owner or operator to achieve maximum final yield (see Sect. 10.2.1).

At present and in the near future, it is very improbable that a fully cost-covering model can be adopted in practice, because the political opposition against it is very strong. Therefore, the combination of at least two of the above mentioned measures is the next best way for achieving a sustainable market introduction of photovoltaics. At least one of the different measures should be oriented to the amount of electricity that is generated by the PV system (see also [133]), and the combination of the different measures should achieve a nearly economical operation of the system.

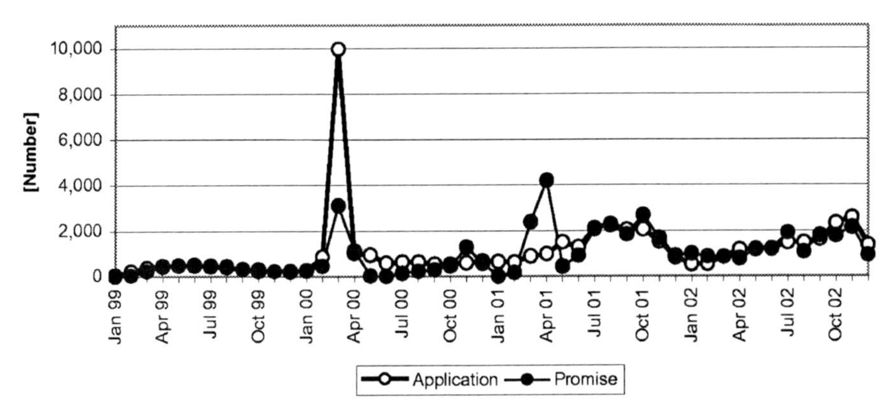

Fig. 11.5. Development of applications and promises (grants) in the 100,000 Roofs Program from January 1999 to December 2002

To prove this statement Fig. 11.5 shows the development of submitted applications and granted loans within the German 100,000 Roof Program. It can be seen very clearly that up to February 2000 the number of applications was low and the number of grants stayed close to the same level. At this time, only the conditions of the 100,000 Roof Program influenced the PV market in Germany. As mentioned above, on 25 February 2000, the new Renewable Energy Law was passed by the German Parliament. The sell-back rate for electricity generated by PV systems was increased to nearly 0.5 € per kilowatt hour.

Figure 11.5 shows that at the same time the number of submitted applications was increasing enormously. From then on, the KfW on behalf the German government changed the conditions for participation in the 100,000 Roof Program repeatedly. Consequently, the number of approved loans decreased. The processing of the applications was delayed for some months, but since May/June 2001 the number of applications and the number of grants moved to the same level.

What can be learned from the trends in Fig. 11.5?

- Only the combination of the two different measures (loan and sell-back rate) made the development of the PV market in Germany sustainable. The number of applications increased and, in addition, the power of the applied respective of the installed PV systems increased from about 2.5 kWp before February 1999 to nearly 4.5 kWp thereafter.
- The measures of supporting the PV market should not be changed during their planned lifetimes.
- The conditions for support measures for market introduction should be in force for a long time.

11.3 Cost of Photovoltaics

11.3.1 Cost of PV Modules

The cost of PV modules depends on materials and technology. We will describe in more detail the following types of PV modules:

- modules based on crystalline silicon,
- modules based on amorphous silicon, and
- thin-film modules with the CIS and CdTe technologies.

Generally, the prices of solar modules per W are governed by a worldwide market, since shipping costs are very low.

Figure 11.6 shows the evolution of the average PV module price in Europe from 1990 to 1999. The average yearly rate of price decrease was 9%.

Figure 11.7 is derived from a general empirical rule that applies for all industrial products, namely, the learning curve. It states that for a certain increase in cumulated production the price drops by a certain fraction. If price is plotted versus cumulated production on a doubly logarithmic scale, a linear function results. Figure 11.7 gives the learning curve for solar modules.

The learning curve is not only a result of mass production, but includes technical progress as well. Mathematically, it can be expressed as:

$$K_x = K_0 \, x^{\frac{\ln(1-L)}{\ln 2}}$$

$$x = \frac{x_t}{x_0} \,,$$

where:
K_x = cost of item (in our case, Wp) produced at time t
K_0 = cost of first item produced
x_t = cumulated production at time t
x_0 = cumulated production at reference time $t = 0$
L = learning rate (e.g., 0.18 or 18%)

From Fig. 11.7 a learning rate of 20% can be derived. This is a rather high factor indicating the large technical and economic potential of this technology. From the learning curve and the estimated yearly production, the future price expectation for PV modules and systems can be calculated. If the present growth rates can be maintained, then the cost goals outlined in the next chapter seem achievable.

Cost of Crystalline Silicon Module

The price of the silicon wafer substantially affects the crystalline silicon module manufacturing cost [149]. Different wafer types, most prominently single crystal or poly crystal, fetch different prices, but also yield different efficiency, as outlined in Chaps. 2 and 3. The cost distribution between wafer, cell technology, and module packaging is about 20 to 40%.

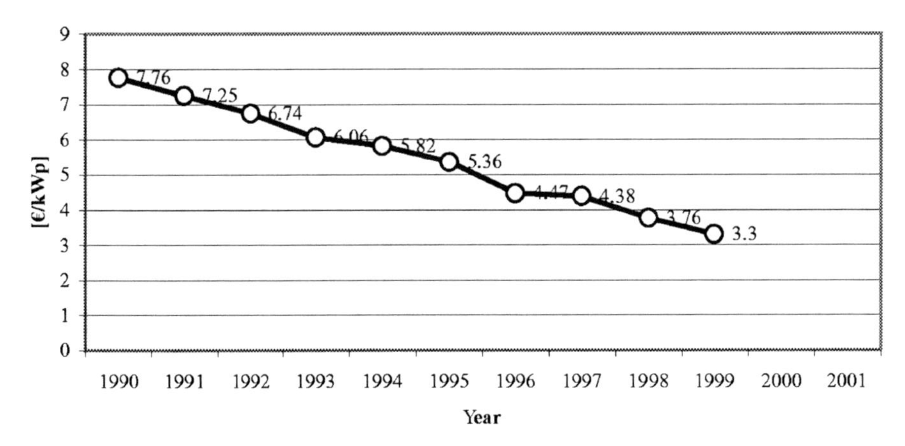

Fig. 11.6. Evolution of the average PV module price in Europe (€/kWp)

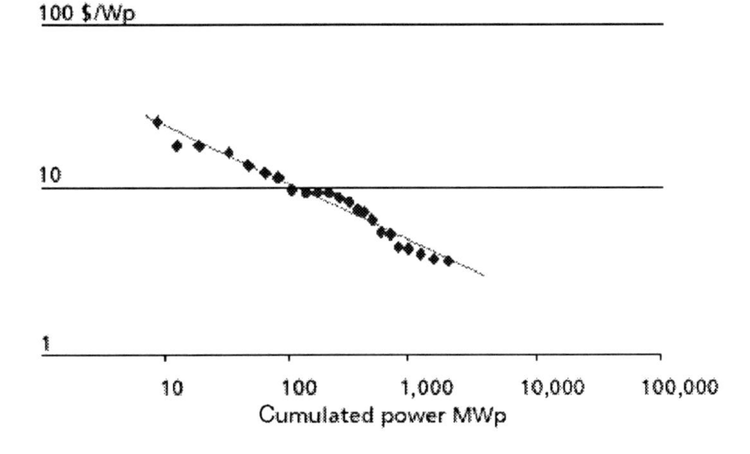

Fig. 11.7. The learning curve: module price versus cumulated production

Cost of Amorphous Silicon Modules

As is reported in [150], the direct manufacturing cost for amorphous silicon modules is around US $2.70 per Wp by a 10 MW plant in 2000. But at present no amorphous silicon operates at this capacity. So a cost comparison with cost of crystalline silicon module is impossible.

Cost of Compound Semiconductor Thin-Film Modules (CIS, CdTe)

At present, only a pilot production of CIS and CdTe thin-film modules exists. This is the reason why usable information about production costs for both kinds of thin-film modules is not available. So a cost comparison of crystalline silicon modules is impossible, but because of the low production volume, it can be assumed that at the production cost of thin-film modules are higher than the prices at which the modules are sold. Market prices are at the same level as crystalline modules on a per watt basis. For more information, see [150].

11.3.2 Cost of PV Systems

There is a big difference between the cost of stand-alone PV systems and the cost of grid-connected PV systems. Therefore, they have to be treated separately.

Cost of Stand-Alone PV Systems

Stand-alone PV systems can have very different purposes and designs. It is not feasible to compare the cost of energy for a solar watch with that of a solar home system. In many cases, stand-alone systems are produced in small numbers, which makes them more expensive. Nevertheless, stand-alone systems are economical, because, in most cases, competing alternative energy sources are even more expensive. So the cost of electricity in stand-alone systems is not very meaningful, but the cost of the modules is a fixed quantity for all applications.

An exception in this consideration are the solar home systems. Mostly, they are produced in great numbers and consist of the same standardized components (see Sect. 7.1). For example, in [149] it is reported that the investment cost for a 50 Wp solar home system with battery and charge controller is about US $500.

Cost of Grid-Connected PV Systems

The cost of complete grid-connected PV systems consists of the cost of:

- PV modules,
- inverter,
- mounting and support frames,
- cabling or electric installation material, and
- planning and installation.

The contribution of the different components to the price of complete grid-connected PV systems depends on the plant size, as is shown in Fig. 11.8. It can vary between 8,570 €/kWp for small plants (1 kWp) and 5,000 €/kWp for large-scale plants (1 MWp). The data used are collected from Germany [151].

Table 11.5 shows the percentage cost contribution of the main components of a PV plant depending on plant size.

From Table 11.3 we see that the percentage of module cost is more than 50% of the total cost of a PV system, and that this percentage is increasing

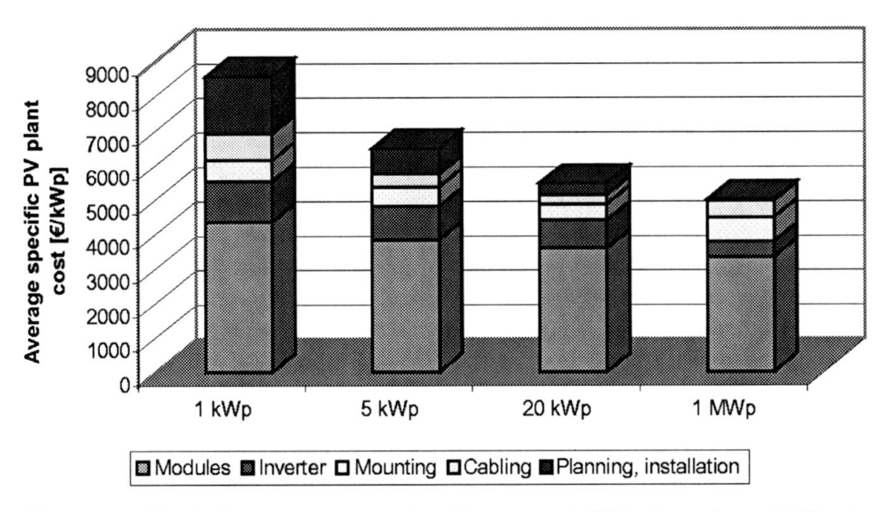

Fig. 11.8. Typical cost structure of grid-connected PV plants from 1 kWp to 1 MWp (prices excluding VAT)

Table 11.5. Percentage of cost contribution of the main components to the average specific cost for grid-connected PV in Germany in 1998 [137]

PV plant size	1 kWp	5 kWp	20 kWp	1 MWp
Modules	50.8	58.9	65.3	66.6
Inverter	13.8	14.9	15.0	9.2
Mounting	7.1	8.6	8.4	13.8
Cabling	8.9	5.9	4.7	9.4
Planning, installation	19.4	11.7	6.6	1.0

with the rated power of the PV system. The cost of the PV module is the main influence on the total cost of PV systems. Table 11.4 shows also that, with the exception of the 1 MWp PV plant, inverter costs for grid-connected PV plants are nearly constant at approximately 15% of the total costs independent of the rated power of the PV system.

Over the past ten years, total costs of PV plants in Germany have decreased by more than a third from approximately 12,800 €/kWp to 7,500 €/kWp [151].

Some reasons for this development are:

- Decreasing cost of PV modules, because of the increase of production and of better production technologies (see Sect. 11.3.1). In addition, the increase of the area of a single PV module – from a half square meter in 1991 to two or three square meters today. This allows quicker installation of the PV generator. Also very important is the increase in efficiency, which decreases the area-dependent costs, as was explained earlier.

- Decreasing cost of the inverter. This came about first by the increase in the number of yearly produced units and, secondly, since 1991, with the launch of serial production of PV inverters, some innovations were introduced in PV inverter topology – for example, transformerless (string) inverters (in 1998), string inverters with higher nominal output (in 1999), multi-string concept (2001), and "Quiet DC Rail" switching concept (in 2001). These innovations lead to the reduction of PV inverter costs and, consequently, to the reduction of PV plant costs [151].
- Many manufacturers have offered so-called PV standard systems to the market. Such PV standard systems consist of a solar generator with a nominal power of one, two, or three kWp, an inverter which is optimally adjusted to the solar generator, and a standardized mounting system for installation on rooftops or on flat roofs (see Sect. 7.2).

All these improvements are reflected by the learning curve (Fig. 11.7), which applies to all industrial products.

11.3.3 Cost of Power Production

The price of PV electricity still exceeds the price of conventional energy and also of most other renewable energies significantly. This is mainly due to the large initial investment in the photovoltaic system. We now describe how the cost of generated electricity can be calculated.

The price of electricity generated by a PV system depends on the following parameters:

- investment cost,
- operating lifetime of the PV system,
- energy output (final yield),
- interest rate for capital, and
- maintenance cost.

Investment cost. The development of the investment cost of PV systems is described in "Cost of Grid-Connected PV System" in Sect. 11.3.2.

Operating lifetime of the PV system. Generally, the lifetime of a PV system is assumed to be twenty-five years. This is the module lifetime guaranteed by most manufacturers; in reality, the lifetime may be longer, but reliable data will only be available in the future.

Energy output (final yield). The final yield of a PV system depends on several parameters, for example, on the quality of the PV system (performance ratio) and the irradiation on the location where the PV system is installed. For more information, see Sect. 10.2.

Interest rate for capital fluctuates from country to country and with time.

Maintenance cost. Normally, a PV system up to a size of 3 or 5 kWp does not need mechanical maintenance. PV systems with a power rating greater than

5 kWp should be controlled periodically every three years. Other possible costs are: insurance, taxes, administration, metering of output, and possibly the leasing of the roof or other space.

The greatest part of the maintenance cost of a grid-connected system is accumulation of a reserve for the possible repair or replacement of the inverter. The maintenance cost is usually calculated at nearly 1% of the investment cost per year.

With the above mentioned parameters the specific costs (Csp) of power production by a PV system can be calculated.

First, the economic viability of the system is calculated. This can be done in different ways, either by the cash equivalent of costs and returns or by the annuity method, since a PV system is a long-term investment that extends far into the future. The cash equivalent method discounts costs and revenue to the present, and if the outcome is positive, then the investment is economic. If we vary the price obtained for the generated electricity, then we obtain the cost of power production for the case where the cash equivalent is zero. How this can be done is shown as an example for discounting interest payments for an investment. The present value CE is:

$$CE = \frac{C}{1+p} + \frac{C}{(1+p)^2} + \ldots = \sum_{i=1}^{n} \frac{C}{(1+p)^i} = \frac{(1+p)^n - 1}{(1+p)^n p} C\,, \quad (11.1)$$

where:
C = initially invested capital,
p = interest rate,
n = lifetime of the plant.

A similar procedure has to be followed for all other positive or negative cash flows.

This is just one possibility to determine the economics of such an investment. Obviously, there are great uncertainties involved in this procedure. Many of the parameters can change over the years, in particular, interest rates, which is the most important one. The value of the generated electricity can be expected to increase over the years. If there is a fixed feed-in rate, then it can be taken as a constant, which makes the calculation more reliable.

It should be stressed that the above procedure is just monetary accounting. Macroeconomic factors like preservation of resources and the environment do not enter here.

Even though the large PV programs and the progress in production of PV components led to a significant cost reduction, grid-connected PV is only economical with support of some kind in middle Europe. Possible future scenarios will be discussed in the following chapter.

12 The Future of PV

12.1 Boundary Conditions for the Future Development of Photovoltaics

12.1.1 Cost Development of Conventional Electricity

The cost development of conventional electricity is a very important boundary condition for the future use of photovoltaics. Generally, it can be assumed that the cost of grid electricity will slowly rise over the next decades. The slope of this rise depends on the following influences: degree of liberalization of electricity markets, margins for the security of supply, proportion of distributed generation, e.g., cogeneration, environmental restrictions, and last but not least, the fate of atomic energy. Practically all these influences, with the exception of liberalization, point toward higher cost of electricity. Another influence comes from the development of storage technologies. Presently, storage of electricity is not a technical, but an economic problem. Large-scale, grid-connected storage is only possible with pumped hydro installations, which are very limited in their potential. Storage batteries are widely used in stand-alone and mobile systems, but are much too expensive for storing grid electricity. If the full potential of photovoltaics is to be realized, better means of storage are necessary. The hydrogen economy would be a solution to the storage problem, as will be pointed out later.

12.1.2 Effects of Liberalization and Environmental Restrictions

The liberalization of the electricity market that is going on worldwide will influence the penetration of grid-connected PV. The following tendencies can be recognized:

- Liberalization will lead to more economics, i.e., lower electricity prices, but this may be limited to large-scale consumers. Therefore, the overall consequence could be neutral relative to the expansion of PV in distributed systems.
- Electricity prices will reflect the real cost of generation. Prices will fluctuate widely depending on the time of day and season. Since a large part of solar electricity is generated during the high price period, this will be beneficial for PV.

– The cost of reserve capacity will lead to a lower security of supply. The role of PV as a back-up source will be favorable for this market.

12.2 Cost and Market Development of Stand-Alone and Grid-Connected Systems

The future of PV depends mainly on the cost development for modules and entire systems. This development can be predicted by extrapolating the learning curve explained in the preceding chapter (Fig. 11.7). This curve establishes a connection between the cumulated (or total installed) capacity in kWp and the price. For modules, which are the most expensive part of a system this curve, has been well established for more than twenty years. It predicts a 20% price reduction for every doubling of cumulated production. Assumptions have to be made about market growth for such a prediction. If present support mechanisms are maintained, the current growth rate will continue into the future. An estimate for the market development is given by W. Hoffmann [152]: 30% per year until 2010 and 25% thereafter. With such assumptions, the price development shown in Fig. 12.1 can be envisioned.

Figure 12.1 gives a possible scenario for grid-connected systems. Cost for distributed PV is plotted for different parts of Europe, insolation of 900 h/year for middle Europe, and 1,800 for southern Europe. This is compared with the utility selling price range for peak power. Also plotted is the cost of bulk power. Since PV power is largely peak power, PV has a chance to become competitive with utility peak power in southern Europe between 2008 and 2020 and around 2030 in northern and middle Europe. Competition with bulk power will take much longer.

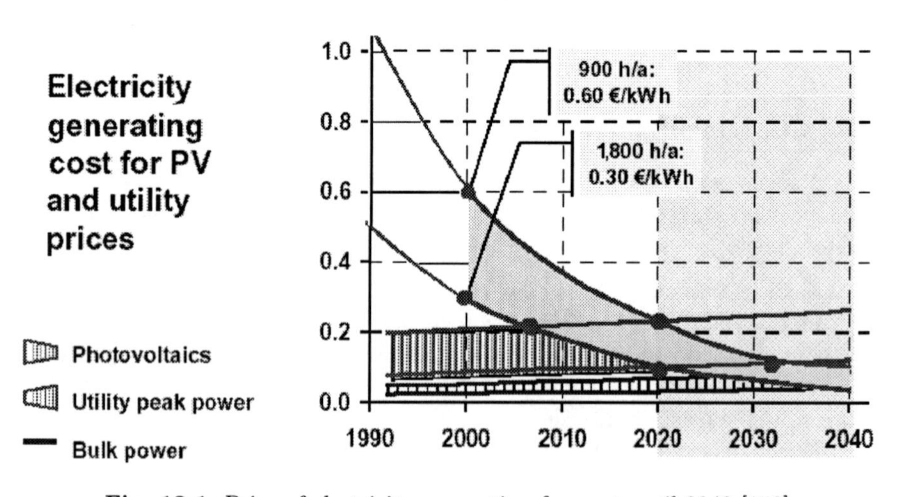

Fig. 12.1. Price of electricity generation forecast until 2040 [152]

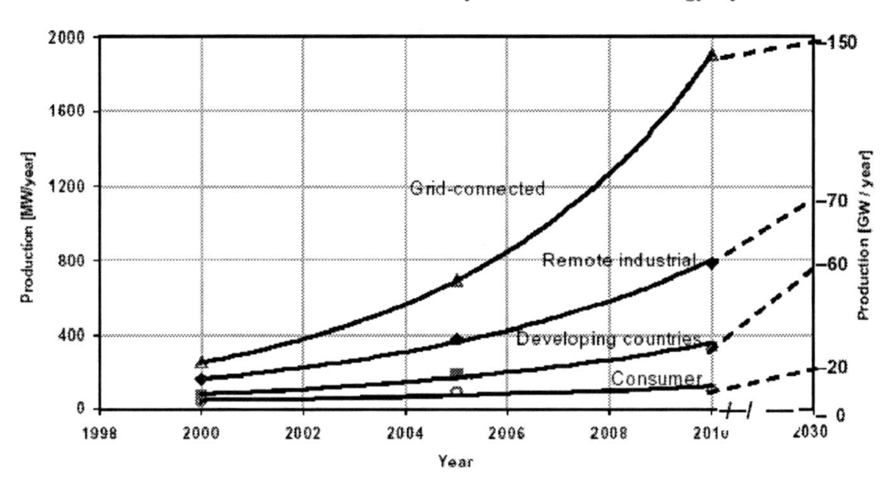

Fig. 12.2. Forecast of the development of different market sectors until 2030. (Note the different scale on the right-hand side of the graph)

Future development of markets: Figure 12.2 gives an impression of the future development of different markets, also from [152]. The grid-connected market will remain the most important sector, but the other sectors, remote industrial, developing countries and consumer products together reach almost the same size. These markets are, of course, interdependent. Only by price reductions due to expanding grid-connected markets can the other sectors grow accordingly.

12.3 PV in a Future Liberalized and Partly Decentralized Energy System

Future electricity grids will have a mixture of central and decentralized generation capacity. Central conventional and renewable plants will cooperate with small and medium distributed generation. Modern electronics will permit a high degree of control, leading to a rapid adjustment of the economic optimum. Very important in this concept is the possibility of local storage of electricity.

12.3.1 Integration of PV into a Decentralized Energy System

Present models of a distributed energy system envision PV at many rooftops or similar structures interacting with an electricity grid and other local generators and consumers. Demand can be adjusted to some degree by central control. Besides PV, other small generators may be connected to the grid like combined heat and power plants, some based on biofuels, fuel cells, and

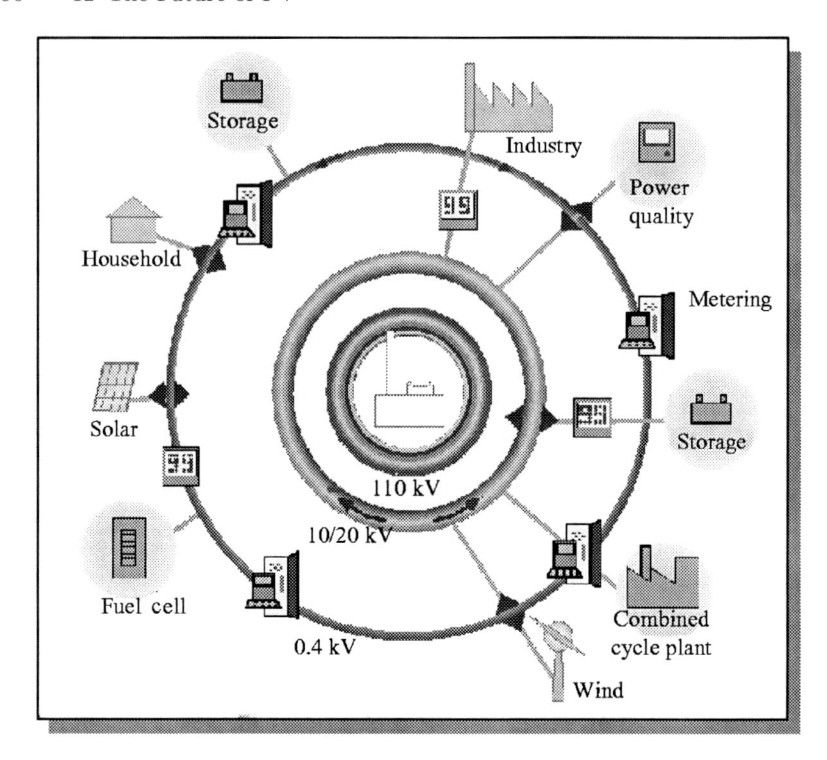

Fig. 12.3. Structure of a future electricity distribution network [153]

local storage in batteries or more likely by hydrogen. Central plants can be planned by weather forecasts predicting wind and sunshine. Different renewable energies acting together have much lower fluctuation than each one by itself. A considerable proportion of local generation and storage may reduce the cost of distribution grids and enhance the security of supply. Much work remains to be done in working out grid and control structures for such a scenario. Undoubtedly, PV will play a major role in such a new electricity scene. The EDIson project supported by the German Ministry of Economics and Technology has the goal of developing and demonstrating modular intelligent systems for such distributed grids. It involves, in particular, the combination of electricity grid and communication systems. The structure of such a future electricity distribution network is shown in Fig. 12.3 [153].

12.3.2 Fully Autonomous Systems, Autonomous House Concepts

A fully autonomous system has already been demonstrated as the Self-Sufficient Solar House (Sect. 7.1.4). In this project, most of the above mentioned techniques were already incorporated and tested. When they will become reality is mainly a question of cost. Fuel cells are under very intense development for mobile applications which are more demanding than stationary ones. If fuel cells become more available, local storage of hydrogen will make distributed systems more independent. From today's point of view, complete autonomy is not a very likely option. But the grid is already in place, so why not use its advantages. If a dwelling with local generation is grid-connected, it can deliver surpluses into the grid and local storage does not have to cover extreme situations. Energy stored locally can be fed into the grid at times of peak power and fetch a higher price. Such a dwelling would have a low heating demand that would be covered by a combined heat and power plant, which could, besides covering local demand, feed energy into the grid when demand is high.

A further scenario developed some time ago recognizes the fact that as energy efficiency in buildings improves the center of demand shifts to mobility. The engine of an automobile is, after all, a combined heat and power plant [154].

12.4 PV in a Centralized Energy System

PV could in a very long-term scenario also play a role in a centralized energy system. Large PV power plants in the multi-megawatt range can be envisioned to produce electricity in forever sunny desert areas or even in outer space. Such scenarios require very low but still possible cost for modules and systems. It is quite clear that these visions can only become reality in the distant future, maybe around the middle of this century. The most important of these schemes will now be described.

12.4.1 Electricity from the Desert

Deserts are regions of abundant sunshine but of very little other use. It is tempting to imagine large-scale solar energy installations in such areas where they are much more efficient than in the areas where most energy is consumed. The big challenge is to transport the energy over large distances to the consumers. Two main techniques can be identified today: high tension power lines and conversion to hydrogen.

Transmission of electricity by power lines over long distances is technically possible today. High voltage dc transmission works with low losses and has been operational for many years. Such lines could be set up, for instance, between Northern Africa and Europe or from the Gobi desert to Japan. An even further reaching idea is to establish a worldwide electricity grid. Then the storage problem could be solved very elegantly, because the sun is always shining in one half of the world. And seasonal differences can also be overcome by energy exchange between the northern and southern hemispheres. Such grids could also be used for wind energy, since many areas with very good wind resources are located in remote parts of the world. The cost of transmission has recently been estimated for transport from north Africa to central Europe [155]. A 5,000 km transmission line has 18% loss at 600 kV and 14% loss at 800 kV. The cost of the losses is estimated at 0.5 cts per kWh and the cost of transmission at 0.5 to 1.0 cts per kWh.

Kurokawa [156] has developed the concept of worldwide electricity grids and recently has initiated a very detailed study on large-scale generation in the Gobi desert with transport to Japan [157]. Generation cost, assuming present technology, was US 17.7 cts/kWh for polycristalline modules and 18.8 cts/kWh for amorphous silicon. The cost of amorphous silicon was higher because of its lower efficiency.

The second option for energy transport is hydrogen. Water is split into hydrogen and oxygen by electrolysis and hydrogen can then serve as the energy carrier. Transport of hydrogen can be accomplished in three different ways:

- Hydrogen is liquefied near a harbor and then transported away by tankers.
- Hydrogen is pressurized and piped to the consumers by pipeline.
- The third pathway is the least explored and its realizability is uncertain: Hydrogen is chemically reacted to form a liquid energy carrier that can be transported, as outlined above, for liquid hydrogen, but does not require refrigeration. At the end of the transport chain, the compound is decomposed and the hydrogen used for energetic purposes. The hydrating reactant can be transported back to the site of electrolysis and reused. Several chemical reactions for the described process have been identified, but no large-scale demonstration has been undertaken so far.

If a hydrogen economy should develop in the future, PV could fit very well into this scenario. A problem is the energy input for electrolysis, refrigeration, and transport of the hydrogen, which makes this pathway less efficient than direct transport of electricity, as described above. However, hydrogen can serve as a storage medium for energy and is also suitable for mobile applications.

As mentioned in Chap. 13, solar thermal plants could have an advantage in large-scale power plants in the desert. Which technology finally wins out depends on which one reaches the cost goals for this application first.

12.4.2 Electricity from Space

The Solar Power Satellite concept was first introduced by Dr. Peter Glaser of Arthur D. Little, Inc., in 1968 [158]. According to this proposal, special satellites with very large areas of photovoltaic modules would be placed in a geostationary orbit. Each satellite would have an area of around 50 km^2. The electricity generated would be transmitted to the surface of the earth by antenna systems using microwave or laser beams. This concept has remained an utopia, although it is taken up periodically. Indeed, it does have attractive features: Sunlight is available in this orbit almost continually and uninterrupted by clouds or atmosphere. Estimates are that a solar cell in space would generate almost eight times more electricity than on earth. Just as large are the obstacles: The cost of lifting large numbers of modules into space appears prohibitive. Also, beaming powerful microwave beams to the earth causes environmental concerns. A newer plan proposes manufacturing solar cells on the moon or on an asteroid in order to avoid the gravity of the earth. Obviously, a solar power satellite is technically possible, but not realizable in the foreseeable future.

13 Other (Perhaps Competing) CO_2-Free Energy Sources

Photovoltaics exists and develops in an interdependent framework of energy sources. In order to judge the future chances of PV, other sources of clean energy have to be considered. In the following chapter, only the most important alternatives for environmentally clean electricity generation are discussed and only as far as necessary to evaluate their future development relative to PV. Many but not all of these sources are renewable and some are in a very early stage of development, which causes some uncertainty in their evaluation.

13.1 Other Renewable Energy Sources

The most important renewable sources of electricity are:

- solar thermal energy,
- hydropower,
- wind energy,
- biomass,
- ocean and wave energy, and
- geothermal energy.

A nonrenewable but nevertheless carbon dioxide-free technology is carbon sequestration, which will also be considered in this context.

Some of these technologies are very compatible with photovoltaics. Others may be competing in the electricity market. In the following chapter, we will give a rough overview of the respective technologies, their state of development, and their potential – as far as possible.

13.1.1 Solar Thermal Energy

The most straightforward conversion of solar radiation is to thermal energy. Besides space and water heating, high-temperature heat can also be converted into electricity just as in conventional plants. Three concepts for thermal electricity generation exist: solar troughs and solar towers for large plants and dishes for decentral application, which will now be discussed. A quite different scheme is the solar tower, which is also under consideration.

Fig. 13.1. View of a single parabolic mirror unit of an advanced SEGS system

Solar Trough Systems

Solar troughs are parabolic trough-shaped mirrors that track the sun. In the focal line an absorber tube filled with the working fluid is arranged. Because of the linear structure, only one-axis tracking is required (Fig. 13.1). The achievable concentration and temperature are lower than for two-axis concentration. High-temperature oil is used as a working fluid in the realized plants, but direct vaporization of water has also been tested. The radiation is concentrated by approximately a factor of 80 and a temperature of up to 400° is reached [159].

Solar trough plants are the only commercial thermal electricity generation technology. In the early 1990s large SEGS (Solar Electricity Generation Systems) with a total power capacity of 354 MW were erected in the Californian Mojave Desert near Dagget and near Kramer Junction (see Fig. 13.2).

Because the tax laws favoring solar electricity were changed, the operating company LUZ had to go out of business soon after the last plant was finished. The existing plants are still successfully operating today, but new plants have not been built since then.

Now trough systems are again in the planning stage in Crete and in Spain. Each of these projects called THESEUS and ANDASOL respectively, has a power of 50 MW. Projects for other systems, in addition, are planned in Egypt, Morocco, and Iran. The total capacity of these projects is projected to reach more than 700 MW. Trough systems are scheduled in the drafts for a solar hydrogen power supply in the future.

Fig. 13.2. Aerial view of the solar power stations SEGS III to VII with an installed capacity of 30 MW each; Kramer Junction (California)

Fig. 13.3. Solar tower power plant – Solar Two at Barstow (U.S.A.)

Solar Tower Plants

Solar tower power plants also work on the principle of concentrated solar radiation. They can only utilize direct radiation. Therefore, they are planned in sun-rich geographic locations like deserts.

Hundreds of rectangular sun-tracked mirrors reflect the solar rays onto a receiver that is located at the top of a tower (Fig. 13.3). The absorber may consist of a system of blackened tubes, but many other designs have also been considered [160]. In the absorber very high temperatures can be generated, of the order of 800°. The heat is transported away by gases, preferably air.

A heat exchanger at the base of the tower transmits the heat energy into a secondary loop. In this loop, water is the working fluid. The rest is

Table 13.1. Solar tower plants [159]

Project	Country	Power MW_{el}	Starting
SSPS	Spain	0.5	1981
EURELIOS	Italy	1	1981
SUNSHINE	Japan	1	1981
Solar One U.S.A.	U.S.A.	10	1982
CESA-1	Spain	1	1983
MSEE/Cat B	U.S.A.	1	1983
THEMIS	France	2.5	1984
SPP-5	Ukraine	5	1896
TSA	Spain	1	1993
Solar Two	U.S.A.	10	1996

a conventional thermal power plant. Several tower concepts contain a heat storage either with the working fluid or molten salt. Thus, a more or less continuous operation during cloudy periods or at night is possible. It is also possible to boost the temperature of the working fluid if it is too low for steam generation. Utilization of the conventional power plant also makes hybrid operation with natural gas or oil possible and permits base load generation.

Thus far, solar tower plants have only been erected as demonstration projects. Such demo plants exist in Spain, Japan, and the U.S. (see Table 13.1). Well-known are the Solar One and Solar Two projects in Barstow both with a power of 10 MW. Solar One consisted of 1,760 heliostats of 40 m^2 each. The entire plant covered an area of 67,000 m^2, the tower was 86 m high. Commercial plants, now in the planning stage, will be much larger, of the order of 10 MW.

Solar thermal power plants play an important role in scenarios for a sustainable world energy supply. They could produce hydrogen by electrolysis that could be transported to industrialized countries.

Solar thermal plants are in direct competition with central PV plants also planned for the distant future. At this time, the achievable costs are distinctly lower for thermal plants, in the range of 0.02 to 0.10 euro/kWh at present and are expected to come down to 0.01 to 0.02 euro/kWh in the future. Development of the thermal technology was very slow in the last decades. A problem is that large projects have to be organized that are difficult to finance. PV, in contrast, has a rapidly developing market, which assures constant cost reductions. It remains to be seen which technology will reach the needed cost goals for central power plants first.

Fig. 13.4. EuroDish-Stirling, Plataforma Solar Almeria, Spain

Dish Systems

Dishes are parabolic mirrors that follow the sun by two-axis tracking. Very high concentration is reached in their focal point. In the focus there is either an absorber like in the solar tower or a Sterling engine [161]. The absorber converts the concentrated radiation into heat and transmits it to a working fluid or gas, which, in turn, can power an electricity generating heat engine. The Sterling engine is a more elegant concept, because it sits directly in the focus and delivers electricity without the need for cumbersome heat transport.

Figure 13.4 shows two so-called EuroDish-Stirling-Systems at the Plataforma Solar Almeria (Spain). They were developed and are operating within a European project for the development of a cost effective 10 kWe solar thermal system for decentralized electric power generation.

Solar Chimney Power Station

A solar chimney power station (Fig. 13.5) combines three well-known technologies: the greenhouse effect, the chimney effect, and an air turbine [162].

The plant consists of three main parts:

- Chimney or tower tube; it is the main characteristic of the solar chimney station. The chimney is up to 1,000 m high and open at the base. Its diameter is 170 m. The large size is necessary, because the efficiency increases with height and the plant becomes more economical.
- Glass roof collector. Around the chimney there is a glass roof collector of a very large area. The present concepts envision an area of several 10,000 m^2. The glass roof collector is open at the outer rim where it is

Fig. 13.5. Solar chimney

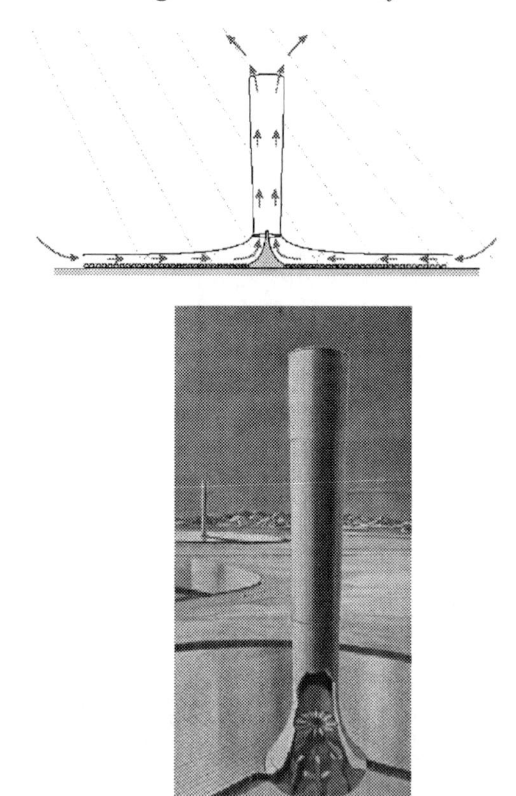

Fig. 13.6. Schematic principle of a solar chimney power station

2–3 m high. The height of the glass roof increases toward the chimney to about 10 m.

– Wind turbine: It is horizontal and sits at the base of the tower.

The operation is very simply explained: The air underneath the glass roof is heated by the greenhouse effect. The temperature difference is about 20 K. The warm air moves toward the chimney and rises up to its top, driving the turbine (see Fig. 13.6).

The overall efficiency of the solar chimney is relatively low, less than 2–3%. Operation can be extended into the evening hours by heat storage in the soil below the glass roof. By covering the ground with water-filled tubes, even twenty-four hour operation can be achieved.

At present, there are plans to erect a commercial plant, 1,000 m high, in Australia. Model calculations indicate a cost of 7.19 ct (€)/kWh [163].

13.1.2 Hydropower

Like wind power, with which we deal in Sect. 13.1.3, hydropower has been used by mankind for a very long time. It is the only renewable energy source that contributes today a sizeable fraction of the energy demand of a number of countries. Almost 20% of the world electricity demand is covered by hydroelectricity. It is estimated that the amount of hydroelectricity produced will grow from 3,000 TWh (Tera Wh = 10^{12} Wh) per year today to about 6,000 TWh per year in the middle of this century, implying a typical growth rate of 2% per year. Typical generation cost is 0.02 to 0.08 euro/kWh. Both the technical potential and the possibilities for further cost reduction are limited. In industrialized countries, many of the more favorable sites are already being exploited.

Table 13.2 shows the technically and economically feasible potential of hydropower generation by region of the world. Because hydropower is already extensively used its additional potential is limited.

Table 13.2. Technically and economically feasible potential of hydropower by region (TWh/year) [164]

Region	Technically feasible potential	Economically feasible potential
Africa	1,750	1,000
Asia	6,800	3,600
North and Central America	1,660	1,000
South America	2,665	1,600

Fig. 13.7. Run-river power station at Ryburg-Schwörstadt (Germany)

There are two kinds of hydropower stations:

- ***Run-of-river power stations*** use the kinetic energy contained in the natural flow of. rivers or creeks. A run-of-river power station does not have a storage capacity worth mentioning. They operate as base load stations. Figure 13.7 shows an example of a run-of-river power station.
- ***Reservoir power station.*** The flowing water is dammed up and stored in a reservoir behind a high (up to 200 m) dam and can be used when required. Reservoir power stations are used to meet the peak load. Some of them can also be used for pumped storage to convert cheap base load surplus into peak load electricity.

The large hydropower plants listed in Table 13.3 are all reservoir plants. They are related to serious damage of the environment. The high dam and the large inundated area (up to 1,000 km^2) are reasons for great concern and will limit the chances for new installations. The problems with the world's largest hydropower plant on the Yangtse river in China are well-known.

Smaller running water power plants (Fig. 13.8) are less problematic. They still have some additional potential.

Hydropower plants can even be combined with PV, as is seen in Fig. 13.9, where the roof of the turbine house is used for a PV generator.

Fig. 13.8. Small hydropower station in Freiburg (Germany)

Fig. 13.9. Water power station Wyhlen (Germany) with a large PV system at the roof of the turbine house

Table 13.3. Large hydropower stations (a selection)

Power station	Operation start	Country	Capacity (MW)
Itaipú	1983	Brazil/Paraguay	12,600
Guri	1986	Venezuela	10,300
Grand Coulee	1942	U.S.A.	6,494
Sajan	1978	Russia	6,400
Krasnojarsk	1967	Russia	6,000
La Grande 2	1979	Canada	5,225
Churchill Falls	1971	Canada	5,225
Bratsk	1961	Russia	4,500

Fig. 13.10. Modern WEC

13.1.3 Wind Energy

Wind is a direct consequence of solar energy. Between 1.5 and 2.5% of solar radiation reaching the earth is converted into wind energy. This is an energy of 2.6 to 4.3×10^3 Terawatt or 1,000 times world electricity consumption.

This potential is not evenly distributed, but depends on geographic location, altitude, and season.

The power density p in W/m^2 of wind is given by the relation

$$p = \frac{1}{2} \sigma_L \, v^3 (W/m^2) , \qquad (13.1)$$

where:
σ_L = density of air, which is constant: 1.2 kg/m^3
v = wind velocity in m/s.

This energy cannot be completely extracted, because it would mean zero velocity at the exit of a wind converter. From aerodynamics it follows that only 59.3% of the energy can be extracted at maximum.

As we can see from (13.1), the energy density of wind depends on the third power of wind velocity. Therefore, the most important parameter for siting windmills is the average wind velocity.

Wind energy was used already in antiquity. It played a big role in Europe in the Middle Ages until it was replaced by fossil energy in modern times.

Today, wind energy is again an important source of renewable energy in many countries. The total installed capacity worldwide was about 40,000 MW at the beginning of 2004, producing about 75 TWh per year. The typical market growth is about 30%, realized at onshore locations. Figure 13.10 shows

Fig. 13.11. Offshore Park Utgrunde in the Swedish part of the North Sea (source: GE Wind Energy)

an example of a modern WEC (Wind Energy Converter). It has a nominal power of 1.75 MW. The rotor diameter is 66 m.

Wind energy is used especially in Germany, the United States, Denmark, and Spain. Development of offshore wind parks has begun (Fig. 13.11).

Typical generation cost is 0.05 to 0.13 euro/kWh and may come down to 0.03 to 0.10 euro/kWh in the long run.

Modern wind energy converters start operating at a wind speed of 4–5 m/s and reach nominal power at 15–17 m/s. At 25 m/s they are turned off to prevent damage.

Wind energy is much further along than PV in terms of market penetration. It has the advantage of much lower cost. Disadvantages are the limitation to high wind locations, the intermittent nature of wind energy, and the problems with public acceptance that are beginning to emerge.

Wind combines very well with PV, because both sources taken together have a higher availability than one of them.

13.1.4 Biomass

Today, biomass contributes significantly to the world energy supply: about 45 EJ per year. Most of this energy comes from the traditional use of biomass.

The modern use of biomass to produce electricity, steam, and biofuels is estimated to be 7 EJ per year. Since the early 1990s, modern biomass has gained considerable interest. Various routes to produce electricity from biomass can be distinguished: combustion of biomass in combined heat and power (CHP) systems, combustion of biomass in stand-alone systems, and co-combustion in power plants. In addition, a relatively young development is of that the biomass integrated gasification/combined cycle (BIG/CC) system, combining the advantage that various fuels can be used and a high electrical efficiency, up to 55% can be achieved. Biomass has a large range of costs depending on the source of biofuel. If waste products can be used, cost can be very low, but bio crops tend to be expensive.

Today, with both state-of-the-art combustion technology and with co-combustion in traditional power plants, a cost of 0.05–0.06 euro/kWh can be obtained. If the BIG/CC technology becomes available, the production cost can drop to about 0.04 euro/kWh.

The potential of biomass is quite large, but cannot be exactly determined, because land area suitable for energy crops is also important for food production. Biomass is a storable source of energy and, therefore, very compatible with photovoltaics.

13.1.5 Ocean and Wave Energy

Wave Energy

Wave energy originates mainly from wind forces, assisted by changes in barometric pressure and tidal forces (see "Tidal Energy" in this section). Details of the formation of ocean waves are very complex, but in a simplified model the water particles move in stationary orbits, forming a transversal wave.

The energy content of waves depends on wind velocity, duration of the wind force, and the distance covered by the wave.

The power of wave energy can be estimated by the following formula.

$$p = \frac{\sigma g^2}{32\pi} H^2 T b , \tag{13.2}$$

where:
H = height of wave
b = width of wave
T = period of wave
σ = density of the seawater
g = gravity constant

Many different schemes for the exploitation of wave energy have been developed in the past. Examples are the Cockerell Raft or the OWC (Oscillating Water Column) techniques. The latter was realized in several prototype installations, "Kaimei," 30 kW in Japan around 1980, and the Kvaerner OWC plant in Norway 1985.

In February 2004, after six years of development, the world's first commercial-scale floating wave energy converter started its operation [165]. The Pelamis machine (Fig. 13.12) measures 120 m long by 3.5 m wide (about the size of four train carriages).

At this point it is difficult to predict the potential, the cost, and the degree of utilization of this energy source.

Fig. 13.12. The Pelamis Offshore Wave Energy Converter (source: www.oceanpd.com)

Tidal Energy

Another possibility of obtaining energy from the ocean is tidal energy, which exploits the different water level between high tide and low tide. The period between two high or low tides is 12.4 h. The tides are caused by gravitational attraction of the sun and moon.

Large tides can be observed in certain coastal formations like narrow inlets or funnel-shaped bays. In such places, the tides can reach 20 m. The tidal power is harnessed by damming the outlets with dams containing reversible turbines. Thus, both directions of the tide can be used. Some tidal power plants exist, for instance, in St. Malo at the mouth of the Rance River in France (240 MW, erected between 1961 and 1967) and at the bay of Fundy in Canada (20 MW, erected in 1984).

The potential of tidal power is limited, because it is restricted to certain coastal formations. In addition, such plants should not have too much environmental impact.

Under consideration of the above mentioned criteria there exist worldwide only forty to fifty potential locations for tidal power plants with a capacity of 200 MW each. At some of those locations, a capacity of more than 1,000 MW should be possible. The total potential capacity is, therefore, 10 GW. The development of tidal energy has been very slow considering the fact that the pilot plants have existed for decades and the technology is basically known.

Ocean Thermal Energy Conversion

This concept uses the temperature difference in tropical oceans between the warm water at the surface and the cold water in the depth of the ocean. It has huge potential, but also severe problems. The biggest handicap is the low Carnot efficiency, because of the relatively small temperature difference that can be exploited.

In one version of OTEC, warm surface water is fed through a heat exchanger in which a liquid with a low boiling point like ammonia is vaporized. The vapor expands and drives a turbine. At a second heat exchanger, which is cooled by cold water pumped from a depth of 700 m, it is liquefied. In another version, the warm surface water is directly vaporized under low pressure and liquefied as before. In both cases, the plants are floating and anchored in deep water.

The first plant of this kind was built off Cuba in 1929 with a power of 22 kW. It operated only for a short time. Forty years later, Japan and the United States began operating OTEC plants again. In 1980, OTEC-1 was in operation off Hawaii. Its efficiency was only 2.5%.

The limited success of the OTEC technique is due to several factors: In addition to the intrinsic low efficiency, a large amount of energy is needed to pump large volumes of water. Plants anchored in the ocean are subject to damage by storms and getting the electricity to shore is a big problem. A possible future solution could be hydrogen technology, but generation of hydrogen by electrolysis and reconversion to electricity further diminishes efficiency.

The potential of ocean thermal energy is very large. An area of 60 million km^2 bounded by 22° latitude on both sides of the equator could generate 10,000 GW, but this is a greatly simplified estimate. So far, not a single commercial plant based on the OTEC principle is in operation.

Use of Marine Currents – Underwater Wind Mills

Currents in the ocean contain large amounts of energy. Compared to wind, the velocity is much lower, but because of the higher mass of water, the energy content is nevertheless high. Almost all properties of wind energy can be transferred to ocean currents with different constants (Fig. 13.13). The dimensions of a turbine in water are much smaller than in air for the same power.

Today, experimental plants are operating in the North Sea. They use tidal currents in relatively shallow water. A 300 kW installation named "Seaflow" is in operation off Lynmouth, North Devon, U.K. It has two 11 m diameter rotors primarily of glass fiber. For 2005, a trial device of 1.3 MW is planned. The developers expect a cost of 6–9 cts (€)/kWh for large farm systems [166]. The energy delivery fluctuates with the tides, because the direction of the current changes every 6.25 hours and is, therefore, very predictable.

It is too early to predict the future importance of energy from ocean currents. The sites for such turbines require high tidal currents.

Fig. 13.13. Seaflow project – view at the twin rotor (source: www.bine.info)

13.1.6 Geothermal Energy

All renewable energy sources mentioned so far (with the exception of tidal energy) are derived from solar radiation. A major potential energy resource not related to solar energy is geothermal energy. This energy originates from the natural cooling of the planet and from radioactive decay. The temperature of the crust of the earth increases with increasing depth. It varies depending on geological conditions, but on average it is 1° per 33 m.

Three types of geothermal resources can be distinguished:

- High temperature steam resources
  Water is heated by geothermal anomalies. Depending on pressure, it can exist either as superheated water or high pressure steam.
- Warm water resources
  These are aquifers with temperatures lower than 100°C. They are not very suitable for electricity generation, but can be very useful for heating.
- Hot dry rock resources
  The hot dry rock technique exploits the heat content of deep rock layers. Whereas the first two resources are relatively close to the surface, hot dry rock heat is tapped several kilometers below the surface.

Fig. 13.14. Twenty plants are still operating at The Geysers (U.S.A.)

Natural Geothermal Energy

The so-called natural geothermal energy exploits the high temperature geothermal resources described above. The power station is mostly located in an area with seismic activities (Fig. 13.14). The steam is tapped by bringing down a bore hole and is fed into a conventional power plant. Very often the emerging steam contains a lot of water, then a separator is needed. A problem is also often the high mineral content of the source.

All geothermal energy power stations that are in operation around the world at present use natural geothermal resources. In Table 13.4 are listed the location and the power capacity.

Artificial Geothermal Energy – Hot Dry Rock

The hot dry rock resource is almost unlimited. Although the hot interior of the earth exists everywhere, some areas are hotter near the surface (geothermal anomalies). These are preferably envisioned for such power plants and counted toward its potential.

The Hot Dry Rock (HDR) technique involves the following steps:

First, a hole is brought down into the anomaly, then water is injected under high pressure, leading to fissures in the rock.

In a predetermined distance to the first hole, a second hole is drilled. Cold water is injected into the first hole, it is heated by the hot rock and is extracted at the second hole. The emerging steam is used to drive a turbine

Table 13.4. Installed geothermal energy power capacity (MWe)

Country	1990	1995	2000
Argentina	0.67	0.67	0
Australia	0	0.17	0.17
China	19.2	28.78	29.17
Costa Rica	0	55	142.5
El Salvador	95	105	161
Ethiopia	0	0	8.52
France (Guadaloupe)	4.2	4.2	4.2
Guatemala	0	33.4	33.4
Iceland	44.6	50	170
Indonesia	144.75	309.75	589.5
Italy	545	631.7	785
Japan	214.6	413.71	546.9
Kenya	45	45	45
Mexico	700	753	755
New Zealand	283.2	286	437
Nicaragua	35	70	70
Philippines	891	1,227	1,909
Portugal (The Azores)	3	5	16
Russia (Kamchatka)	11	11	23
Thailand	0.3	0.3	0.3
Turkey	20.6	20.4	20.4
U.S.A.	2,774.6	2,816.7	2,228
Total	5,831.72	6,833.38	7,974.06

operating with a low boiling point working fluid to produce electricity. At this time, only prototypes are in operation.

The potential of HDR is very high and resources exist in almost all countries. Only for Germany has a potential for the entire base load been determined. The cost of energy is estimated to 18–20 cts/kWh right now, but can reach 6–8 cts/kWh in the long run.

Figure 13.15 shows the Fenton Hill HDR Test Site at Los Alamos, New Mexico. It is the first HDR test location in the world. Research and development activities are also carried out at Bad Urach (Germany) and Soultz-sous-Forêts (France).

Geothermal energy typically delivers base load electricity and, therefore, is very compatible with the peak load aspects of photovoltaics.

Fig. 13.15. Fenton Hill HDR Test Site

13.2 Carbon-Free Combustion of Fossil Fuels: Carbon Sequestration

13.2.1 What Is Carbon Sequestration?

If it were possible to burn fossil fuels, in particular, coal, without carbon dioxide emission, this type of energy conversion would become acceptable concerning its climatic effect. It would still not be a renewable resource, because fossil fuel reserves are limited, but, in particular, coal would be available maybe for the rest of this century.

Carbon sequestration means that the CO_2 from the combustion of carbon-containing fuel is captured and somehow stored such that it cannot escape into the atmosphere. Capturing CO_2 is not an easy task, because it has to be separated from the flue gases, which normally contain mostly nitrogen plus many other, partly noxious combustion products. CO_2 is a very stable compound that does not easily react with other chemicals.

Just as difficult is storing it in gaseous or solid form, but some possibilities exist, as will be outlined here.

Another aspect is the energy balance. The capture and storage of CO_2 requires energy input. Therefore, the efficiency of the power plant will be reduced.

If carbon sequestration could be realized, the present energy supply system could largely remain unchanged and more time would be available to convert to other energy sources. It is not surprising that some nations with large coal reserves like the U.S. have initiated research programs to explore carbon sequestration. It should be stressed that work on this topic is in a very early stage and it cannot be predicted if these techniques will be realistic and what their cost would be. The U.S. carbon sequestration program has very ambitious cost goals. Whereas present cost is estimated to be US $100–300 per ton of avoided carbon emission, the goal is to reduce this (as a proof of concept) to $10 or less by 2015. This means that the cost of energy services should not be increased by more than 10%. It remains to be seen if this milestone can be achieved. Past experience has taught us that the cost of new energy technologies is always drastically underestimated at the beginning of development.

13.2.2 CO_2 Capture and Separation

Carbon sequestration consists of two steps: separation of CO_2 from the other exhaust gases and the actual sequestration, i.e., permanent storage of the CO_2.

Capture: Several possibilities exist to separate CO_2.

- Oxygen combustion. If only oxygen is used for the combustion, the reaction product of carbon combustion is pure CO_2; no further separation is necessary.
- Pre-combustion decarbonization. Syngas from an oxygen-fired gasifier is shifted to provide a mixture of H_2 and CO_2 at high pressure. Glycol is used to capture CO_2.
- Post-combustion capture. Many possibilities: advanced amine absorption, physical sorbents, CO_2 selective membranes, sorbent membranes.

Storage (sequestration): The task is to find means to store the CO_2 safely for an indeterminate time and to assure that it does not escape into the atmosphere. An important boundary condition is that no adverse effects on the environment should be connected with the storage. The most important options are listed here:

- Depleted oil reserves. CO_2 is already today pumped into depleted oil fields to improve oil productivity, but not with the intention to store CO_2 permanently. Storage in underground formations is today the most realistic pathway. One could imagine that CO_2 pipelines transport the gas back to depleted oil and natural gas pipelines. The cost would be comparable to the present cost of transporting and recovering natural gas. Still, the cost of energy would be significantly higher.
- Unminable coal beds. Coal beds can absorb great amounts of CO_2, while at the same time methane is liberated, which can serve as an H_2-rich fuel source.
- Saline formations. Such formations can also store CO_2. This technique is the only practically tested type of carbon sequestration. At the Sliepner natural gas field in the Norwegian part of the North Sea, 1 million tons of CO_2 is injected into a saline formation per year.

Table 13.5. Carbon sequestration potentials

	Gt C
Forests and soils	> 100
Geological formations	300–3,200
Oceans	1,400–20,000,000

Source: US-DOE Carbon sequestration overview

- Terrestrial uptake. The fixing of CO_2 by plant growth is also sometimes mentioned as a means for carbon sequestration, but this a very limited reservoir because the amount of additional biomass in the ecosphere is limited.
- Ocean injection. At a greater depth in the ocean, CO_2 forms hydrates, because of the high pressure, and potentially remains there. This type of storage has the greatest potential, but also the greatest uncertainties and potential dangers. Big question marks are the stability of storage and the environmental impact of increased CO_2 concentrations in the deep ocean.

An overview of the most important carbon sinks is listed in Table 13.5.

To put these figures in perspective, we consider the world consumption of coal and gas, which presently is about 8 Gigatons per year. As already stated, forests and soils have only limited capacity, but geological formations could probably take up the emissions of most of the existing fossil reserves. Ocean storage, despite its high potential, appears at present unrealistic.

In summary: Carbon sequestration is a possible option to prolong the life of the present carbon-based energy supply. The advantage is that not much would have to be changed, but realization depends on the solution of many yet unknown technical and environmental problems.

14 Popular Killing Arguments Against PV and Why They Are Not Valid

In this last chapter we summarize the contents of this book by listing popular arguments against PV and show why they are wrong. If one gives popular talks about PV one is invariably confronted with these arguments that are partly just prejudices but partly also serious objections that have to be discussed. From the end of 2002 into the beginning of 2003 *Photon* published a series of articles on exactly this topic. This chapter picks up this approach referring to the appropriate previous chapters for more detail.

14.1 Solar Modules Consume More Energy for Their Production Than They Ever Generate

This argument is heard most frequently and is just as incorrect as it is widespread. It is treated in detail in Sect. 9.3. There, we show that the energy payback time ranges from 7.3 years for monocrystalline silicon to 1.5 years for thin-film modules (Table 9.1). These figures are from 1998. Newer studies for entire systems show energy payback times from 7 to 2.8 years (Table 9.2). This shows that module and systems technologies are not static, but are constantly improving. The above payback times have to be seen in the light of the fact that all crystalline silicon modules have lifetime guarantees of twenty-five years and thin-film technologies of ten years. The actual lifetime can be expected to be considerably higher. Payback time can be expected to shrink with decreasing module and system cost, because improvements in manufacturing usually involve less energy consumption. On the other hand, energy is only a minor contribution to the cost of modules and so is not as important as, for instance, labor. Also, if recycling of modules is practiced, the payback decreases further. These figures apply to Central Europe. In sunnier climates, these times are, of course, considerably less.

In comparing renewable energy systems with conventional plants it should also be remembered that fossil and even nuclear plants do not have a payback time, because they constantly need input of primary energy. They only convert primary energy into electricity.

14.2 PV Produces More Greenhouse Gases Than It Saves

A similar calculation as for energy can be carried out for greenhouse gases: How long does it take to recover the greenhouse gases that are emitted during manufacture of the modules and other components of a PV system? The figures vary greatly from one investigation to another. Some researchers ended up with very high values for greenhouse emission during the operation of PV systems. It turned out that they included fossil back-up plants (see also Sect. 14.3) of the same power as the PV system in the balance and took an unfavorable energy mix for the production of silicon. Most of the raw silicon, however, is produced with clean hydropower in countries like Norway. In a more realistic scenario calculated by Photon [167], greenhouse gas recovery times range between 2.4 and 3.6 years for the energy mix of Central Europe.

14.3 Grid-Connected PV Requires Lots of Back-Up Fossil Power Plants

A very effective killer argument is: Solar cells produce energy only when the sun is shining. During the night and in cloudy periods back-up power is required. Therefore, PV does not replace power plants, only primary energy. In reality, the following points change the situation:

- Demand is strongly fluctuating and, therefore, reserve capacity is needed anyway. Studies have shown that a 10–20% contribution of PV in relation to total capacity can be accommodated easily by the grid. Today, we are far from this value.
- PV energy production coincides largely with peak demand.
- In a larger grid, local fluctuations are dampened.
- PV is not the only renewable energy source in a future grid. Other sources like wind have a different and partly opposite time dependence than PV.
- Storage, local or central, will become available in the future. At present, only pumped hydro storage can accept surpluses.
- Demand control will become more widespread in the future. Demand can then be adapted to the availability of solar energy.

Many more concepts can be imagined for the future. An example is adapting the production of energy-rich products to the availability of energy and storing a product like aluminum.

14.4 PV Is Too Expensive

PV electricity is indeed expensive, but this is just a transitory situation. Admittedly, PV is the most expensive source of renewable energy. Grid-connected systems generate electricity for about 50 € cts/kWh in Central

Europe today, but prices are dropping. In more favorable locations, cost is only half as high. The learning curve (Fig. 11.7) is well established and will continue depending on market growth in the next centuries. It will still take decades of continued support for grid-connected systems before prices drop to a competitive level. Why should we incur all these expenses? PV is the only renewable energy source that can be used everywhere on the globe. It has almost unlimited potential, it is long-lived, reliable, and does not harm the environment; if installed on buildings, it does not consume area. In the long term, it will be an indispensable part of the energy system in almost every country. From Fig. 12.1, which shows a possible scenario of price development, we see that PV will first become competitive with peak power, but prices will continue to drop beyond this point.

14.5 PV Is Not Ready for Marketing, More Research Is Required

The history of the PV market and of technology development is a good example of how improvement of a product is stimulated by interaction with markets. In the beginning, only consumer products were equipped with solar cells. Later, stand-alone and remote applications became attractive. These markets are today fully economic. The real goal is, however, to utilize the large potential of PV for world electricity supply, which requires very low prices. These prices are technically feasible if markets grow much faster than normal development would permit. Past experience has shown that a new source of energy cannot compete initially with established energy sources. Nuclear energy even today receives subsidies. It should not be neglected that even fossil energy is subsidized in many countries. Critics say that since the cost of PV is so high at present we should not push it into the market but rather spend more for research. But this would not work. Although more research is certainly required, the lowering of cost can only materialize if there is feedback from markets. This is demonstrated by the present rapid growth of subsidized grid-connected markets, which has led to exactly the price reductions predicted by the learning curve.

14.6 Installation of PV in the Northern Half of Europe Does Not Make Sense Because the Same Solar Cells Generate Electricity Much Cheaper in the South

It is certainly true that a PV module can generate around twice as much energy in a very sunny climate than in Central or Northern Europe or in similar climates in America or Asia. Nevertheless, it is attractive to install grid-connected systems in these locations because:

- Energy arriving at the premises of the customer is much more expensive than at the central power plant. The cost to the customer includes transmission and metering, transmission losses, administration, and taxes. Thus, this cost is five to six times higher than the generating cost.
- Solar energy generated in the south and transported over large distances incurs larger transmission cost and transmission losses. Its cost has to be competitive with the cost at the power plant.
- Countries in the sun belt in Africa and Asia tend to be politically unstable. The risk of energy production in these countries is reduced if there is also generation in Europe.

14.7 PV Involves Toxic Materials

This point has been treated in Sect. 9.1. There is general agreement that PV modules do not have any negative effects during operation. There remains the question of manufacturing. Crystalline silicon involves diverse chemicals and treatments during manufacture and technology, as described in Sect. 3.1. Some of these chemicals are toxic or flammable, but the technology is well controlled based on decades of experience from the chip industry. The modules themselves contain mainly silicon and glass with traces of lead used for soldering. Amorphous silicon modules are equally harmless. Some thin-film cells like CdTe indeed contain toxic elements. Their manufacture has been shown to be without problems, but questions can be asked about environmental effects if those modules are damaged or destroyed by fire. Studies have shown that the amount of Cd in those modules is so small that no significant increase of Cd concentration in the environment results, even in such an unlikely event. Future recycling of modules will further reduce environmental impact, as described in Sect. 9.5.

14.8 PV Consumes Valuable Land Area

Presently, large PV plants in the megawatt range are being installed, especially in Germany. The reason is the high feed-in tariff intended for roof integration, which makes large-scale installations profitable because of their lower cost. The new feed-in law recognizes this by a significantly lower tariff for such plants. However, it is not intended to eliminate this market completely, because the industry feels it needs these projects to sustain market growth. Although the large projects use land that is not used like former refuse dumps or, in one case, a former army camp, resistance to such systems is rising. It should be stressed that for the foreseeable future PV does not need these spaces, because there is still a huge reserve of unused roof, facade, and sound barrier space.

If agricultural land should really be considered for PV, it is even possible to continue its use for crops, as we have shown in Sects. 8.4 and 9.4. In desert areas, it is even possible to provide shading by modules that at the same time generate electricity for irrigation and desalination of water.

14.9 PV Competes for Roof Space with Thermal Collectors

Without question, PV modules and solar thermal collectors utilize the same southern roof space, but nonetheless no competition is to be expected. First of all, there is enough roof space available to accommodate both of them. Some considerations further mitigate this competition.

- Thermal collectors need only limited roof space because they have a very high efficiency ($\sim 50\%$). If they are used for water heating, an accepted ratio is $1\,m^2$ per person. Normal roofs are much larger, and the rest can be used for PV. The situation is different if the collectors are used for space heating. In this case large collectors are required, but in summertime those collectors are stand idle, in contrast to PV, which always produces useful energy.
- Thermal collectors, in contrast to PV, are insensitive to partial shading, so they can be placed in locations that are not suitable for modules, and in this way, the available roof space can be used more efficiently.

In laboratories today, work is also underway to develop hybrid collectors that generate both electricity and heat. In those collectors, the solar modules serve simultaneously as absorbers. Crystalline silicon cells are not particularly practical for this purpose because with increasing temperature their efficiency drops, but if higher bandgap semiconductors should become available this problem will be solved. Hybrid collectors could become the most elegant way to solve the roof space problem, which in any case is not of much concern today.

14.10 A Feed-in Tariff Causes Unacceptably High Electricity Cost

In Chap. 11, it was shown that feed-in tariffs are the most effective way to support the rapid development of the PV market. In such an arrangement, the added cost is included in the energy cost of the customer. In this manner, two advantages result: No government subsidy is needed, so political uncertainties about continued support can be avoided. Secondly, the cost is allocated to the source of the problem, namely, energy consumption. Since utilities like to justify price increases with the heavy burden due to renewable energies,

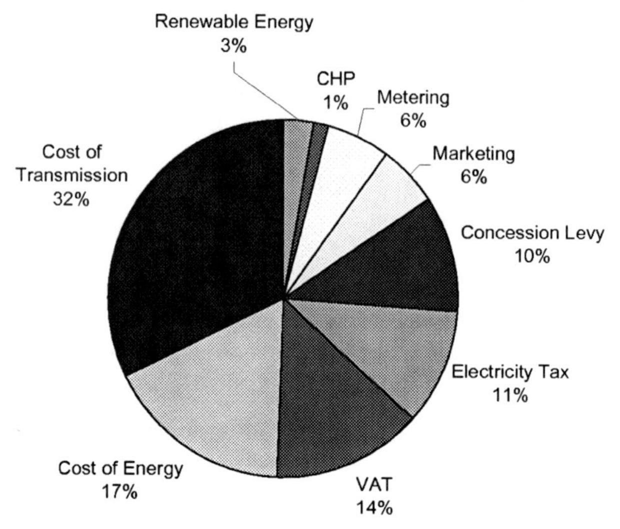

Fig. 14.1. Composition of electricity cost in Germany. Total cost is 19.2 cts

it is necessary to look at the composition of electricity cost. In Germany, where experience with feed-in tariffs exist, the data are well-known. Does this lead to intolerably high electricity prices? No! The data for the largest European PV market, Germany, are available. Right now, PV adds less than 0.1 cts/kWh of electricity cost. The total contribution of renewable energies is 0.52 cts/kWh mainly due to wind. But how will it develop in the future with the constantly increasing capacity of PV and other renewable energies. The answers can be found in a recent study undertaken for the German Ministry for the Environment by J. Nitsch and coworkers [168]. In a scenario investigated by this group, the maximum additional cost for renewable energies will reach 0.8 cts/kWh in 2015, it will drop to zero in 2040, and save money after that. PV will then still need support in the feed-in model, but it can be expected that long before this PV electricity will be more economic if used directly by the consumer, as pointed out in Sect. 14.6.

Of course, the total support for PV is still large (around 1,400 Mio €/year at the maximum) because the energy market is enormous, but the study shows that the introduction of renewable energies provides a cheaper and more reliable supply in the long run if the initial investment is undertaken now. The cost is not prohibitive, namely, 2% of the total energy market on average between 2000 and 2020.

References

1. Becquerel AE, Comt Rend. Academie d. Sciences **9** (1839) p. 561
2. Chapin DM, Fuller CS, Pearson GL, J. Appl. Phys. **25** (1954) p. 676
3. Dietl D, Helmreich, Sirtl E, *Crystals: Growth, Properties and Applications*, Vol. 5, Springer (1981) p. 57
4. Staebler DL, Wronski CR, Appl. Phys. Lett. **31** (1977) p. 292
5. Erge T et al., The German experience with grid-connected PV-systems, Solar Energy **70**, No. 6 (2001) p. 479–487
6. Becker H et al., Five years of operational experience in the German 1000-Roofs-PV Programme – Results of monitoring and system inspection, Proc. 14th European Photovoltaic Solar Energy Conf., Barcelona (1997) p. 1677
7. Gamberale M, Castello S, Li Causi S, The Italian Roof-Top Program: Status and perspectives, PV in Europe, Rome, Italy (2002) p. 1012–1015
8. Wilk, H, Grid-connected PV-Systems in Austria, lessons learned, Proc. 17th European Photovoltaic Solar Energy Conf., Munich (2001) p. 907
9. www.solarch.ch – Solarstromstatistik
10. Kaltschmitt M, Wiese A, *Erneuerbare Energieträger in Deutschland, Potentiale und Kosten*, Springer (1993)
11. Nitsch J, Trieb F, *Potenziale und Perspektiven regenerativer Energieträger*, Study for the German Parliament (March 2000)
12. Goetzberger A, Kleiss G, Reiche K, Nordmann T, Proc. 2nd World Conf. on Photovoltaic Energy Conversion, Vienna, Austria (1998) p. 3481
13. Goetzberger A, Knobloch J, Voss B, *Crystalline Silicon Solar Cells*, John Wiley & Sons (1998)
14. Palm J, Lerchenberger A, Kusian W, Krühler W, Endrös A, Mihalik G, Fickett B, Nickerson, Jester T, Proc. 16th European Photovoltaic Solar Energy Conf., Glasgow (2000) p. 1222
15. Dietl J, Helmreich D, Sirtl E, *Crystals: Growth, Properties and Applications*, Vol. 5, Springer (1981) p. 57
16. Goetzberger A, Shockley W, J. Appl. Phys. **31** (1960) p. 409
17. Goetzberger A, William Cherry Award Lecture, Proc. 26th IEEE Photovoltaic Specialists Conf., Anaheim, USA (1997) p. 1
18. Block HD, Wagner G, Proc. 16th European Photovoltaic Solar Energy Conf., Glasgow (2000) p. 1059
19. Nussbaumer H, Herstellung und Eigenschaften rekristallisierter Siliziumschichten, Dissertation, University of Konstanz (1996)
20. Ciszek TF, J. Crystal Growth **66** (1984) p. 655
21. Eyer A, Räuber A, Goetzberger A, Optoelectronics-Devices and Technologies **5**, No. 2 (1990) p. 239

22. Bergmann RB, Werner JH, Kristallzüchtung für die Photovoltaik – Forschung in Deutschland, Mitteilungsblatt der Deutschen Gesellschaft für Kristallwachstum und Kristallzüchtung **59** (1994) p. 15
23. Kalejs JP, Mackintosh BH, Sachs EM, Wald FV, Proc. 14th IEEE Photovoltaic Specialists Conf., San Diego (1980) p. 13
24. Wald F. In: Dietl J, Helmreich D, Sirtl E, *Crystals, Growth, Properties and Applications*, Vol. 5, Springer (1981) p. 157
25. Schmidt W, Woesten B, Proc. 16th European Photovoltaic Solar Energy Conf., Glasgow (2000) p. 1083
26. Mackintosh BH, Quellette MP, Piwczyk BP, Rosenblum, MD, Kalejs JP, Proc. 28th IEEE Photovoltaic Specialists Conf., Anchorage, Alaska, USA (2000) p. 46
27. Schmidt W, Woesten B, Proc. 16th European Photovoltaic Solar Energy Conf., Glasgow (2000) p. 1083
28. Sachs EM, Ely D, Serdy J, Edge Stabilized Ribbon (ESR) growth of silicon for low cost photovoltaics, J. Crystal Growth **82** (1987) p. 117–121
29. Ciszek TF, Hurd JL, Schietzelt MJ, J. Electrochem. Soc. **129** (12), (1982) p. 2823
30. Ciszek TF. In: *Silicon Processing for Photovoltaics*, ed. by Khattak CP and Ravi KV, Elsevier, Amsterdam (1985) p. 131
31. Wallace RL, Hanoka JI, Narasimha S, Kamra S, Rohatgi A, Proc. 26th IEEE Photovoltaic Specialists Conf., Anaheim, USA (1997) p. 1277
32. Janoch R, Wallace RL, Hanoka JI, Proc. 26th IEEE Photovoltaic Specialists Conf., Anaheim (1997) p. 95
33. Gabor AM, Hutton DL, Hanoka JI, Proc. 16th European Photovoltaic Solar Energy Conf., Glasgow (2000) p. 2083
34. Green MA et al., Prog. Photovolt. Res. Appl. **7** (1999) p. 31
35. Knobloch J, Glunz SW, Biro D, Warta W, Schäffer E, Wettling W, Proc. 25th IEEE Photovoltaic Specialists Conf., Hawaii, USA (1994) p. 1477
36. Green MA, High efficiency solar cells, Trans. Tech. Publications (1987) p. 170
37. Wenham SR, Honsberg CB, Green MA, Sol. En. Mat. Solar Cells **34** (1994) p. 110
38. Lammert MD, Schwartz RJ, IEEE TED **31** (1977) p. 337
39. Swanson RM et al., IEEE TED **31** (1984) p. 661
40. Sinton RA et al., IEEE Electron Dev. Lett. **7** (1986) p. 567
41. Hezel R, Sol. En. Mat. Solar Cells **74** (2002) p. 25
42. Sakata H, Kawamoto K, Taguchi M, Baba T, Tsuge S, Uchihashi K, Nakamura N, Kiyama S, Proc. 28th IEEE Photovoltaic Specialists Conf., Anchorage, Alaska, USA (2000) p. 7
43. Gee JM, Schubert K, Basore PA, Proc. 23rd IEEE Photovoltaic Specialists Conf., Louisville, USA (1993) p. 265
44. Van Kerschaver E et al., Proc. 2nd World Conf. on Photovoltaic Energy Conversion, Vienna, Austria (1998) p. 1479
45. Schneiderlöchner E, Preu R, Lüdemann R, Glunz SW, Prog. Photovolt. Res. Appl. **10** (2002) p. 29
46. Goetzberger A, Hebling C, Schock H-W, Materials Science and Engineering **40** (2003) p. 1–46
47. Wolf M, Proc. 14th IEEE Photovoltaic Specialists Conf., San Diego (1980) p. 674

48. Spitzer M, Shewchun J, Vera ES, Loferski JJ, Proc. 14th IEEE Photovoltaic Specialists Conf., San Diego (1980) p. 375
49. Goetzberger A, Proc. 15th IEEE Photovoltaic Specialists Conf., Kissimmee (1981) p. 867
50. Goetzberger A, Knobloch J, Voss B, Techn. Digest of 1st International Photovoltaic Science and Engineering Conf., Tokyo (1984) p. 517
51. Hebling C, Eyer A, Faller FR, Hurrle A, Lüdemann R, Reber S, Wettling W, Festkörperprobleme, Advances in Solid State Physics **38** (1998) p. 607
52. Bergmann RB, Appl. Phys. A **69** (1999) p. 187–194
53. Yamamoto K et al., Proc. 26th IEEE Photovoltaic Specialists Conf., Anaheim, USA (1997) p. 575
54. Green MA, Zhao J, Zheng G, Proc. 14th European Photovoltaic Solar Energy Conf., Barcelona (1997) p. 2324
55. Basore PA, Proc. 29th IEEE Photovoltaic Specialists Conf., New Orleans, Lousiana (2002) p. 49
56. Zimmermann W et al., Proc. 16th European Photovoltaic Solar Energy Conf., Glasgow (2000) p. 1144
57. Lüdemann R et al., Proc. 26th IEEE Photovoltaic Specialists Conf., Anaheim, USA (1997) p. 159
58. Takami A, Proc. 13th European Photovoltaic Solar Energy Conf., Nice, France (1995) p. 59
59. Barnett AM, Prog. Photovolt. Res. Appl. **5** (1997) p. 317
60. Brendel R, *Thin-Film Crystalline Silicon Solar Cells*, Wiley-VCH, Weinheim (2003)
61. Catchpole KR, Weber KJ, Sproul AB, Blakers AW, Proc. 2nd World Conf. on Photovoltaic Solar Energy Conversion, Vienna, Austria (1998) p. 1336
62. Hamamoto S et al., Proc. 14th European Photovoltaic Solar Energy Conf., Barcelona (1997) p. 2328
63. Tayanaka H, Yamaguchi Matsushita KT, 2nd World Conf. on Photovoltaic Solar Energy Conversion, Vienna, Austria (1998) p. 1272
64. Rinke TJ, Bergmann RB, Werner JH, Proc. 16th European Photovoltaic Solar Energy Conf., Glasgow (2000) p. 1128
65. Matsushita T et. al., Proc. 16th European Photovoltaic Solar Energy Conf., Glasgow (2000) p. 1679
66. Goetzberger A, German Patent 10311893 (2002)
67. Honsberg C et al., Proc. 17th European Photovoltaic Solar Energy Conf., Munich (2001) p. 3
68. Shockley W, Queisser HJ, Detailed balance limit of efficiency of pn junction solar cells, J. Appl. Phys. **32** (1961) p. 510
69. Chittick RC, Alexander JH, Sterling HF, J. Electrochem. Soc. **116** (1969) p. 77
70. Spear WE, LeComber PG, J. Non-Crystalline Solids **8-10** (1972) p. 727
71. Carlson D, Wronski C, Appl. Phys. Lett. **28** (1976) p. 671
72. Staebler DL, Wronski CR, Appl. Phys. Lett. **31** (1977) p. 292
73. Tawada Y, Yamagishi H, Techn. Digest of 11th Intern. International Photovoltaic Science and Engineering Conf., Sapporo (1999) p. 3
74. Meier H et al., Proc. MRS Symp., Spring Meeting, San Francisco, Vol. 507 (1998) p. 139–144
75. Wagner S, Shay JL, Migliorato P, Kasper HM, Appl. Phys. Lett. **25** (1974) p. 434

76. Kazmerski LL, White FR, Morgan GK, Appl. Phys. Lett. **46** (1976) p. 268
77. Dimmler B, Schock HW, Progr. Photovolt. Res. Appl. **4** (1996) p. 425–433
78. Contreras MA et al., Progr. Photovolt. Res. Appl. **7** (1999) p. 311
79. Cusano DA, Solid-State Electronics **6** (1963) p. 217
80. Tyan Y, Perez-Albuerne EA, Proc. 16th IEEE Photovoltaic Specialists Conf., San Diego (1982) p. 794
81. Ferekides C et al., Proc. 23rd IEEE Photovoltaic Specialists Conf., Louisville, USA (1993) p. 389
82. Bett AW et al., Proc. 29th IEEE Photovoltaic Specialists Conf., New Orleans, Lousiana (2002) p. 844
83. O'Regan B, Grätzel M, Nature **252** (1991) p. 737
84. Meissner D, Photon International **34** (1999) p. 34
85. Sariciftci NS et al., Science **258** (1992) p. 1474
86. Kolodinski S, Appl. Phys. Lett. **63** (1993) p. 2405
87. Werner JH et al., Proc. 13th European Photovoltaic Solar Energy Conf., Nice, France (1995) p. 111
88. Würfel P, Sol. En. Mat. Solar Cells **46** (1997) p. 43
89. Luque A, Martí A, Phys. Rev. Lett. **78** (1997) p. 5014
90. Martí A, Proc. 16th European Photovoltaic Solar Energy Conf., Glasgow (2000) p. 15
91. Goetzberger A, Proc. 17th European Photovoltaic Solar Energy Conf., Munich (2001) p. 9
92. Course book for the seminar "Photovoltaic Systems," prepared as a part of the Comett project SUNRISE issued in 1995 by Fraunhofer Institute for Solar Energy Systems, Freiburg, Germany
93. Maycock P, PV market update, Renewable Energy World (July–August 2003) p. 84–89
94. Adib R, Rural electrification: A new financial product for microfinance institutions – Rural electricity loans, 3rd Annual Seminar on New Development Finance, Germany (1999)
95. Lund P, Improved stand alone PV systems with fuel cells, Proc. 17th European Photovoltaic Solar Energy Conf., Munich (2001) p. 2422–2425
96. Mayer D et al., Photovoltaic/electrolyzer/fuel cell hybrid system the tomorrow power station for remote areas, Proc. 17th European Photovoltaic Solar Energy Conf., Munich (2001) p. 2539–2542
97. Goetzberger A, Stahl W, Bopp G, Heinzel A, Voss K, Advances in Solar Energy **9** (1994) p. 1
98. Goetzberger A, Bopp G, Grießhaber W, Stahl W, The PV/hydrogen/oxygen-system of the Self-Sufficient solar house Freiburg, Proc. 23rd IEEE Photovoltaic Specialists Conf., Louisville, USA (1993) p. 1152–1158
99. Surendra TS et al., Solar PV water pumping comes of age in India, Proc. 29th IEEE Photovoltaic Specialists Conf., New Orleans, Lousiana (2002) p. 1485–1488
100. Sastry EVR, Indian National Programme on solar photovoltaics, Proc. 12th International Photovoltaic Science and Engineering Conf., Jeju, Korea (2001) p. 821–824
101. Broker C et al., Design of a photovoltaically operated reverse osmosis plant in off-grid operation for desalination of brackish water, Proc. 26th IEEE Photovoltaic Specialists Conf., New Orleans, Lousiana (2002) p. 1329–1332

102. Voss K et al., Building energy concepts with photovoltaics – Concept and examples from Germany, Advances in Solar Enery (2002) p. 235
103. Meinhardt M, Cramer, G, Cost reduction of PV-inverters – Targets, pathways and limits, Proc. 17th European Photovoltaic Solar Energy Conf., Munich (2001) p. 2410–2413
104. Myrzik JMA, Calais M, String and module integrated inverters for single-phase grid connected photvoltaic systems – A review, IEEE Powertech, Bologna (2003)
105. Goetzberger A, Walze G, Proc. 14th International Photovoltaic Solar Energy Conf., Bangkok, Thailand (2004) p. 719
106. Gutschner M, Nowak S, Ruoss D, Toggweiler P, Potential for building integrated photovoltaics. PV in Europe, Rome, Italy (2002) p. 944–947
107. Sick F, Erge T, *Photovoltaics in Buildings – A Design Handbook for Architects and Engineers*, James & James, London (1996)
108. Goetzberger A, Nordmann T, Frölich A, Kleiss G, Hille G, Reise C, Wiemken E, van Dijk V, Betcke J, Pearsall N, Hynes K, Gaiddon B, Castello S, The potential of PV noise barrier technoloy in Europe, Proc. 16th European Photovoltaic Solar Energy Conf., Glasgow (2000) p. 2912
109. Evaluation of the potential of PV noise barrier technology for electricity production and market share. Final report thermie project: EUPVNBPOT, Project Number SME-1479-97, report period 8.6.98–7.6.99
110. Nordmann T, Frölich A, Dürr M, Goetzberger A, First experience with a bifacial PV noise barrier, Proc. 16th European Photovoltaic Solar Energy Conf., Glasgow (2000) p. 1777–1782
111. Goetzberger A et al., "Sun farming" or how to reconcile PV and nature, Proc. 2nd World Conf. on Photovoltaic Solar Energy Conversion, Vienna, Austria (1998) p. 3481–3485
112. Goetzberger A, Zastrow A, On the coexistenc of solar-energy conversion and plant cultivation, Int. J. Solar Energy **1** (1982) p. 55–69
113. Dobon F et al., Very low concentration system (VLC), Proc. 17th European Photovoltaic Solar Energy Conf., Munich (2001) p. 664–667
114. Mohring HD, Gabler H, Solar electric concentrators with small concentration ratios: Field experience and new developments, Proc. 29th IEEE Photovoltaic Specialists Conf., New Orleans, Lousiana (2002) p. 1608–1611
115. Sala G, Antón I et al., The EUCLIDES-THERMIE concentrator power plant in continous operation, Proc. 17th European Photovoltaic Solar Energy Conf., Munich (2001) p. 488–491
116. Ortabasi U et al., DISH/Photovoltaic Cavity Converter (PVCC) system for ultimate solar-to-electricity conversion efficiency – General concept and first performance predictions, Proc. 29th IEEE Photovoltaic Specialists Conf., New Orleans, Lousiana (2002) p. 1576–1579
117. Garboushian V, Roubideaux D, Johnston P, Hayden H, Initial results from 300 kW high-concentration PV installation, Proc. 12th International Photovoltaic Science and Engineering Conf., Jeju, Korea (2001) p. 103–106
118. Frankl P, Life cycle assessment (LCA) of PV systems – And overview and future outlook, PV in Europe, Rome, Italy (2002) p. 588–592
119. Möller J, Heinemann D, Wolters D, Proc. 2nd World Conf. on Photovoltaic Energy Conversion, Vienna, Austria (1998) p. 2279
120. Bernreuter J, Mehrfache Ernte, Photon (December 2002) p. 49–51

121. Warburg N et al., Recycling of photovoltaic systems – Aims and roads, Proc. 17th European Photovoltaic Solar Energy Conf., Munich (2001) p. 600–601
122. *Photovoltaikanlagen – Untersuchungen zur Umweltverträglichkeit*, BINE Projekt Info-Service Nr. 6, Fachinformationszentrum Karlsruhe, September (1998)
123. Wambach K, *Untersuchungen zu den technischen Möglichkeiten der Verwertung und des Recyling von Solarmodulen auf der Basis von kristallinem und amorphem Silicium*, PILKINGTON Solar International GmbH, Köln (1998)
124. Bernreuter J, Recycling im Backofen, Photon (August 2003) p. 38–30
125. Menezes S, Li Y, Menezes SJ, Closed-loop approach for recycling CdTe PV-modules, Proc. 17th European Photovoltaic Solar Energy Conf., Munich (2001) p. 478–483
126. Hoffmann W, PV solar electricity: One among the new Millenium industries, Proc. 17th European Photovoltaic Solar Energy Conf., Munich (2001) p. 851–861
127. Jahn U, Nasse W, Performance analysis and reliability of grid-connected PV systems in IEA countries, Proc. 3rd World Conf. on Photovoltaic Energy Conversion, Osaka, Japan (2003) p. 2148–2151
128. Jahn U, Performance, reliability and user experiences. In: *Practical Handbook of Photovoltaics: Fundamentals and Applications*, ed. by Markvart T and Castaner L, Elsevier (2003)
129. Reise C et al., Remote performance check for grid connected PV system using satellite data, Proc. 16th European Photovoltaic Solar Energy Conf., Glasgow (2000) p. 2618–2621
130. Quintana MA, King DL, McMahon TJ, Osterwald CR, Commonly observed degradation in field-aged photovoltaic modules, Proc. 29th IEEE Photovoltaic Specialists Conf., New Orleans, Lousiana (2002) p. 1436–1439
131. Reis AM, Coleman NT, Marshall MW, Lehman PA, Chamberlin CE, Comparison of PV module performance before and after 11-years of field exposure, Proc. 29th IEEE Photovoltaic Specialists Conf., New Orleans, Lousiana (2002) p. 1432–1435
132. Laukamp H, The new german electric safety standard for residential PV systems, Proc. 25th IEEE Photovoltaic Specialists Conf., Washington DC (1996) p. 1406–1408
133. Gabler H, Heidler K, Hoffmann VU, Market introduction of grid-connected photovoltaic installations in Germany, 2nd World Conf. on Photovoltaic Solar Energy Conversion, Vienna, Austria (1998) p. 3413–3417
134. Tampier M, Effective green power policies, Refocus – The International Renewable Energy Magazine (Jan/Feb 2003) p. 30–33
135. Measuring and Monitoring Approaches, IEA PVPSII/3 Internal working document 21/10.98
136. www.iea-pvps.org
137. Scientific final report 1000 Roof Program (in German), unpublished
138. Haas R et al., *Soziologische Begleitforschung zum Österreichischen 200-kW-Photovoltaic-Breitentest*, TU Wien, Vienna (March 1997)
139. Nordmann T et al., Proc. 16th European Photovoltaic Solar Energy Conf., Glasgow (2000) p. 2784
140. Spitaleri C, Seminara A, Ceci C, "Sun at School", The Italian PV School Programm, PV in Europe, Rome, Italy (2002) p. 1032–1034

141. Close, J, The Hong Kong Schools Solar Education Programm, Proc. 12th International Photovoltaic Science and Engineering Conf., Jeju, Korea (2001) p. 817–818
142. Osborn DE, Collier DE, Utility business opportunities with the commercialization of grid-connected photovoltaic systems: SMUD as a case example, Proc. 12th International Photovoltaic Science and Engineering Conf., Jeju, Korea (2001) p. 379–387
143. Martin G, O'Toole M, Greening the city, Refocus – The International Renewable Energy Magazine (Jan/Feb 2003) p. 20–24
144. Sayigh A, Photovoltaic status in the United Kingdom, Proc. 12th International Photovoltaic Science and Engineering Conf., Jeju, Korea (2001) p. 653–662
145. Hoffmann VU, Heins M, Photovoltaik-Anlagen im DBU-Förderprogramm "Kirchengemeinden für die Sonnenenergie" – Erste Ergebnisse und Erfahrungen, 17. Symposium Photovoltaische Solarenergie, Kloster Banz, Bad Staffelstein, Germany (2002) p. 191–196
146. Nordmann T, Subsidies versus rate based incentives; for technology-, economocal- and market-development of PV. The European experience, Proc. 3rd World Conf. on Photovoltaic Energy Conversion., Osaka, Japan (2003) p. 2544–2547
147. www.solarstromag.net
148. Sawin JL, Charting a new energy future, Solar Today (March/April 2003) p. 37–39
149. Hoffmann W, PV solar electricity: One among the new Millenium industries, Proc. 17th European Photovoltaic Solar Energy Conf., Munich (2001) p. 851–861
150. Jones E, Frantzis L, Wood M, Lee C, Wormser P opportunities for cost reduction in photovoltaic modules, Proc. 16th European Photovoltaic Solar Energy Conf., Glasgow (2000) p. 2100
151. Meinhardt M, Cramer G, Cost reduction of PV-inverters – Targets, pathways and limits, Proc. 17th European Photovoltaic Solar Energy Conf., Munich (2001) p. 2410–2413
152. Hoffmann W, PV solar electricity industry: Market growth and perspective, Proc. 14th International Photovoltaic Solar Energy Conf., Bangkok, Thailand (2004) paper PL-8
153. Schmidt H, Weissmüller G, Stefanblome T, Povh D, Intelligent energy distribution networks through the use of innovative decentralised generation, storage and communication systems – EDISon, Proc. 17th European Photovoltaic Solar Energy Conf., Munich (2001) p. 962–963
154. Goetzberger A, 9th Symposium Photovoltaische Solarenergie, Kloster Banz, Bad Staffelstein, Germany (1994) p. 337
155. Quaschning V, Sone Wind & Wärme **5** (2004) p. 23
156. Kurokawa K, Proc. 14th International Photovoltaic Solar Energy Conf., Bangkok, Thailand (2004) p. 359
157. Ito M et al., Proc. 14th International Photovoltaic Solar Energy Conf., Bangkok, Thailand (2004) p. 667
158. Potter S, Solar power satellites – An idea whose time has come. A www.page by Rich Brown – www.freemars.org/history/sps.html
159. Geyer M et al., Parabolrinnensysteme. In: *Themenheft 2002: Solare Kraftwerke*, ForschungsVerbund Sonnenergie, Berlin, Germany (2003) p. 14

160. Pitz-Paal R, Buck R, Hoffschmidt B, Solarturmkraftwerkssysteme. In: *Themenheft 2002: Solare Kraftwerke*, ForschungsVerbund Sonnenergie, Berlin, Germany (2003) p. 23
161. Laing D, Schiel W, Heller P, Dish-Stirling-Systeme. In: *Themenheft 2002: Solare Kraftwerke*, ForschungsVerbund Sonnenergie, Berlin, Germany (2003) p. 30
162. www.sbd.de – The Solar Chimney
163. Schlaich J, Aufwindkraftwerke. In: *Themenheft 2002: Solare Kraftwerke*, ForschungsVerbund Sonnenergie, Berlin, Germany (2003) p. 85
164. *Hydropower and the World's Energy Future*. International Hydropower Association, Ontario, Canada (2000); www.icold-cigb.org/PDF/Whitepaper.pdf
165. Ocean Power Delivery Ltd., Edinburgh – Press Release (23rd February 2004)
166. Refocus – The International Renewable Energy Magazine (March/April 2004) p. 46
167. Photon **1** (2003) p. 51 – first of a series of articles
168. Nitsch J et al., Ecologically Optimized Extension of Renewable Energy Utilization, BMU-Brochure, Federal Ministry for the Environment, Nature Conservation and Nuclear Safety, Berlin, Germany (2004); www.erneuerbare-energien.de

Index

Springer Series in

OPTICAL SCIENCES

Volume 1

1 **Solid-State Laser Engineering**
By W. Koechner, 5th revised and updated ed. 1999, 472 figs., 55 tabs., XII, 746 pages

Published titles since volume 80

80 **Optical Properties of Photonic Crystals**
By K. Sakoda, 2nd ed., 2004, 107 figs., 29 tabs., XIV, 255 pages

81 **Photonic Analog-to-Digital Conversion**
By B.L. Shoop, 2001, 259 figs., 11 tabs., XIV, 330 pages

82 **Spatial Solitons**
By S. Trillo, W.E. Torruellas (Eds), 2001, 194 figs., 7 tabs., XX, 454 pages

83 **Nonimaging Fresnel Lenses**
Design and Performance of Solar Concentrators
By R. Leutz, A. Suzuki, 2001, 139 figs., 44 tabs., XII, 272 pages

84 **Nano-Optics**
By S. Kawata, M. Ohtsu, M. Irie (Eds.), 2002, 258 figs., 2 tabs., XVI, 321 pages

85 **Sensing with Terahertz Radiation**
By D. Mittleman (Ed.), 2003, 207 figs., 14 tabs., XVI, 337 pages

86 **Progress in Nano-Electro-Optics I**
Basics and Theory of Near-Field Optics
By M. Ohtsu (Ed.), 2003, 118 figs., XIV, 161 pages

87 **Optical Imaging and Microscopy**
Techniques and Advanced Systems
By P. Török, F.-J. Kao (Eds.), 2003, 260 figs., XVII, 395 pages

88 **Optical Interference Coatings**
By N. Kaiser, H.K. Pulker (Eds.), 2003, 203 figs., 50 tabs., XVI, 504 pages

89 **Progress in Nano-Electro-Optics II**
Novel Devices and Atom Manipulation
By M. Ohtsu (Ed.), 2003, 115 figs., XIII, 188 pages

90/1 **Raman Amplifiers for Telecommunications 1**
Physical Principles
By M.N. Islam (Ed.), 2004, 488 figs., XXVIII, 328 pages

90/2 **Raman Amplifiers for Telecommunications 2**
Sub-Systems and Systems
By M.N. Islam (Ed.), 2004, 278 figs., XXVIII, 420 pages

91 **Optical Super Resolution**
By Z. Zalevsky, D. Mendlovic, 2004, 164 figs., XVIII, 232 pages

92 **UV-Visible Reflection Spectroscopy of Liquids**
By J.A. Räty, K.-E. Peiponen, T. Asakura, 2004, 131 figs., XII, 219 pages

93 **Fundamentals of Semiconductor Lasers**
By T. Numai, 2004, 166 figs., XII, 264 pages

94 **Photonic Crystals**
Physics, Fabrication and Applications
By K. Inoue, K. Ohtaka (Eds.), 2004, 209 figs., XV, 320 pages

95 **Ultrafast Optics IV**
Selected Contributions to the 4th International Conference
on Ultrafast Optics, Vienna, Austria
By F. Krausz, G. Korn, P. Corkum, I.A. Walmsley (Eds.), 2004, 281 figs., XIV, 506 pages

Springer Series in

OPTICAL SCIENCES

96 **Progress in Nano-Electro Optics III**
Industrial Applications and Dynamics of the Nano-Optical System
By M. Ohtsu (Ed.), 2004, 186 figs., 8 tabs., XIV, 224 pages

97 **Microoptics**
From Technology to Applications
By J. Jahns, K.-H. Brenner, 2004, 303 figs., XI, 335 pages

98 **X-Ray Optics**
High-Energy-Resolution Applications
By Y. Shvyd'ko, 2004, 181 figs., XIV, 404 pages

99 **Few-Cycle Photonics and Optical Scanning Tunneling Microscopy**
Route to Femtosecond Ångstrom Technology
By M. Yamashita, H. Shigekawa, R. Morita (Eds.) 2005, 241 figs., XX, 393 pages

100 **Quantum Interference and Coherence**
Theory and Experiments
By Z. Ficek and S. Swain, 2005, 178 figs., approx. 432 pages

101 **Polarization Optics in Telecommunications**
By J. Damask, 2005, 110 figs., XVI, 528 pages

102 **Lidar**
Range-Resolved Optical Remote Sensing of the Atmosphere
By C. Weitkamp (Ed.), 161 figs., approx. 416 pages

103 **Optical Fiber Fusion Splicing**
By A.D. Yablon, 2005, 100 figs., approx. IX, 310 pages

104 **Optoelectronics of Molecules and Polymers**
By A. Moliton, 2005, 200 figs., approx. 460 pages

105 **Solid-State Random Lasers**
By M. Noginov, 2005, 149 figs., approx. XII, 380 pages

106 **Coherent Sources of XUV Radiation**
Soft X-Ray Lasers and High-Order Harmonic Generation
By P. Jaeglé, 2005, 150 figs., approx. 264 pages

107 **Optical Frequency-Modulated Continuous-Wave (FMCW) Interferometry**
By J. Zheng, 2005, 137 figs., approx. 250 pages

108 **Laser Resonators and Beam Propagation**
Fundamentals, Advanced Concepts and Applications
By N. Hodgson and H. Weber, 2005, 497 figs., approx. 790 pages

109 **Progress in Nano-Electro Optics IV**
Characterization of Nano-Optical Materials and Optical Near-Field Interactions
By M. Ohtsu (Ed.), 2005, 127 figs., approx. XIV, 225 pages

110 **Kramers–Kronig Relations in Optical Materials Research**
By V. Lucarini, K.-E. Peiponen, J.J. Saarinen, E.M. Vartiainen, 2005, 37 figs., approx. IX, 170 pages

111 **Semiconductor Lasers**
Stability, Instability and Chaos
By J. Ohtsubo, 2005, 169 figs., approx. XII, 438 pages

112 **Photovoltaic Solar Energy Generation**
By A. Goetzberger and V.U. Hoffmann, 2005, 138 figs., approx. IX, 245 pages

Printing: Strauss GmbH, Mörlenbach
Binding: Schäffer, Grünstadt

Printed in the United States
144852LV00003B/27/A

9 783540 236764